Evan J. Nicholas
2 Perkins Ave, Apt. #4
Narragansett, R.I.
02882
(401) 789-9905

Folk Management in the World's Fisheries: Lessons for Modern Fisheries Management

Folk Management in the World's Fisheries

Lessons for Modern Fisheries Management

Christopher L. Dyer *and* James R. McGoodwin,
Editors

UNIVERSITY PRESS OF COLORADO

Published by the University Press of Colorado, P.O. Box 849,
Niwot, Colorado 80544

The University Press of Colorado is a cooperative publishing enterprise supported, in part,
by Adams State College, Colorado State University, Fort Lewis College, Mesa State College,
Metropolitan State College of Denver, University of Colorado, University of Northern
Colorado, University of Southern Colorado, and Western State College of Colorado.

Library of Congress Cataloging-in-Publication Data

Folk management in the world's fisheries: lessons for modern
   fisheries management / Christopher L. Dyer and James R.
   McGoodwin, editors.
      p. cm.
   Includes bibliographical references and index.
   ISBN 0-87081-325-0
   1. Fishery management — Citizen participation. 2. Fishery
management — Social aspects. I. Dyer, Christopher L., 1954– .
II. McGoodwin, James R.
SH329.C57F65 1994
333.95'6 — dc20   94-1669
                                      CIP

The paper used in this publication meets the minimum requirements of the American
National Standard for Information Sciences — Permanence of Paper for Printed Library
Materials. ANSI Z39.48–1984

∞

10 9 8 7 6 5 4 3 2 1

Contents

Contributors

Eugene N. Anderson, a cultural anthropologist at the University of California, Riverside, is one of the foremost scholars in the study of folk knowledge. He is the author of *The Floating World of Castle Peak Bay.*

C. Gaye Burpee is a Ph.D. candidate in the Department of Crop and Soil Sciences, Michigan State University.

Christopher L. Dyer is a research evaluations specialist with Aguirre International in Washington, D.C.

Lawrence F. Felt is a sociologist at the Memorial University, Newfoundland, and specializes in human dimensions of maritime fisheries.

Duane A. Gill, a sociologist with the Social Science Research Center, Mississippi State University, specializes in disaster research, human ecology, and fisheries management.

David B. Halmo is an applied anthropologist at the Bureau of Applied Research in Anthropology, University of Arizona, and author of numerous articles in applied anthropology.

Svein Jentoft is a distinguished research scientist with the Institute of Social Science, University of Tromso, and a leading authority on fisheries policy and co-management in Norway.

R. E. Johannes, research scientist with CSIRO Marine Laboratories, Hobart, Tasmania, and author of numerous articles and monographs, pioneered studies of indigenous resource conservation and is an expert regarding indigenous resource practices in Oceania.

Richard L. Leard is a fishery biologist and fisheries management plan coordinator with the Gulf States Marine Fisheries Commission. He is a former director of the Department of Natural Resources, Marine Resource Division, for the State of Mississippi.

James R. McGoodwin, a professor of anthropology at the University of Colorado, Boulder, is a pioneer in the study of human dimensions of contemporary fisheries issues and is the author of *Crisis in the World's Fisheries: People, Problems, and Policies.*

Knut H. Mikalsen is a distinguished research scientist at the Institute of Social Science, University of Tromso, and an expert on issues of contemporary fisheries management in Norway.

Craig T. Palmer is an anthropologist at the University of Colorado, Denver, specializing in the maritime anthropology of New England and Newfoundland.

Evelyn W. Pinkerton, a research associate with the School of Community and Regional Planning, University of British Columbia, is a leading authority on co-management in fisheries and is editor of the book *Co-Operative Management of Local Fisheries: Directions for Improved Management and Community Development.*

Caroline Pomeroy holds a postdoctoral fellowship at the Workshop in Political Theory and Policy Analysis, Indiana University, and did her doctoral research on the fisheries of Lake Chapala, Mexico.

Kenneth Ruddle, an independent research consultant, is a leading expert on the fisheries of the Pacific Rim and is co-editor with Tomoya Akimichi of *Maritime Institutions in the Western Pacific.*

Brent W. Stoffle is an applied anthropologist working for the National Park Service, Washington, D.C., specializing in natural and cultural resource management.

Richard W. Stoffle is an eminent applied anthropologist associated with the Bureau of Applied Research in Anthropology, University of Arizona, and is the author of *Caribbean Fishermen-Farmers.*

William Ward recently graduated with a B.A. in history from the University of Houston, Clear Lake, and is pursuing further work concerning maritime peoples.

Priscilla Weeks, a cultural anthropologist and research associate of the Environmental Institute of Houston, is co-editor with Richard Pollnac of *Mariculture in Developing Countries: Problems and Perspectives.*

Foreword

Fisheries social scientists probably need little encouragement to read this collection of articles because many of the authors are already well-known to them. This foreword is therefore concerned mainly with the value of the book to fisheries biologists, who are generally less familiar with its subject matter.

Implicit in the arguments of fisheries social scientists is the assumption that the ultimate objective of fisheries management should be to sustain human societies in general and fishers in particular. Many fisheries biologists have agreed — in principle. All too often, however, they have added, "But management that does not sustain fish cannot sustain fishers; our job is to sustain fish."

This is a cop-out. Sustaining fish and sustaining fishers cannot generally be treated as independent objectives — a point to which fisheries social scientists as well as fisheries biologists do not always manifest a working commitment. To reformulate a statement of one of the contributors to this book, it seems to be human nature, and thus characteristic of both social scientists and biologists, to construct a picture based on partial information and then defend it energetically.

For historical reasons, biologists are far more influential than social scientists in fisheries management. Already beleaguered for decades by economists demanding more influence on decision making, fisheries biologists are now confronted increasingly by social scientists making similar demands. This makes them nervous, and conventional fisheries biology curricula have not prepared them for it (as Ward and Weeks point out in Chapter 4).

But if they are to carry out their responsibilities as fisheries managers properly, biologists must either develop necessary social science expertise or draft into their ranks those who have it. Either course of action will require that fisheries biologists gain greater familiarity with the rapidly expanding social-science literature on fisheries.

This book is not a bad place to start. Its contributors analyze a wide range of examples of the involvement of fishers in the management of their fisheries, and they do not shrink from describing the attendant problems, as well as the advantages, of such arrangements. Just as fisheries biologists have progressed beyond assuming that simple equilibrium yield models and tidy spawning/recruitment relationships underlie fish population dynamics, fisheries social scientists are discovering that some of their comfortable earlier assumptions about the virtues of folk management (also called co-management or cooperative management) were oversimplified.

For biologists unwilling to tackle this book in its entirety, I suggest that they read the "Lessons for Modern Fisheries Management" found at the end of each chapter, as well as the excellent "Summary and Conclusions" by Evelyn Pinkerton. She puts the observations of the other contributors into a frame of reference that should appeal to fisheries biologists, given their predilection for classification and ordination, and that should stimulate them to read further. They may also find Ward and Weeks's chapter especially interesting. Rather than studying fishers, Ward and Weeks set out to find what makes some fisheries managers tick, and reach some provocative conclusions.

R. E. JOHANNES
Hobart, Tasmania, Australia

Introduction

Christopher L. Dyer and James R. McGoodwin

Folk management in the fisheries is management by and for fishing people themselves. It naturally arises as an inevitable outcome of resource utilization by fishing peoples. Formally defined, it is any localized behavior originating outside state control that facilitates the sustainable utilization of renewable natural resources. We take a broad view, feeling it can be unconscious, intentional, or inadvertent, and can be directed toward controlling human use strategies (e.g., limiting entry) or enhancing the productivity of the resource (e.g., seeding an oyster bed).

Some may find it problematic that we include unconscious, unintentional, and inadvertent practices under the rubric of folk management. Hames (1987:93), for instance, asserts that "any persuasive account of conservation as a human adaptation requires a theory that shows that conservation is by design . . . and not a side-effect of some other process." We fundamentally disagree, however, noting that localized societies develop as ongoing, adaptive experiments and that local peoples are often unconscious of the antecedent, trial-and-error experiences that led to their institution of certain core traditions. Indeed, once positively adaptive behaviors become customary and are described by local peoples as "just how we do things around here," they may persist for a long time thereafter without any conscious effort to make them persist.

In the current literature on fishing peoples, *folk management* has a number of synonyms, including *traditional management, localized management, self-management, indigenous management, community-based management, organic management,* and *bottom-up management* (cf. Bailey 1988, Cordell 1989, Johannes 1978, Klee 1980, McCay 1981, McCay and Acheson 1987, Ruddle and Akimichi

1984, and Ruddle and Johannes 1985). Recently it has been discussed under the rubric of "traditional sea tenure," which Ruddle and Akimichi describe as "the ways in which fishermen perceive, define, delimit, 'own,' and defend their rights to inshore fishing grounds" (1984:1). However, we are sure that it often entails much more than merely asserting rights to inshore fishing grounds and may include strategies for limiting fishing effort and enhancing the productivity of fish stocks. Thus, it often differs little in practice from biologically based, modern fisheries management.

Not long ago there were few published studies concerning folk management and therefore little information about it available to modern fisheries managers. This began to change in the 1970s, and by the 1980s there was a veritable boom under way in studies of localized management, such that by now, as Ruddle and Akimichi observe, it is "one of the most significant 'discoveries' to emerge from the last ten years of research in maritime anthropology," even if it is "nothing new to fishermen" (1984:1). Lamentably, and despite the availability of this information, there has been little utilization of it by fisheries managers.

More than ten years ago, McCay (1981:5–6) described three basic strategies that are common in folk management: (1) the assertion of property rights over prime fishing spaces, (2) the exclusion of outsiders from fishing areas that fishing communities assert are their own, and (3) the manipulation of information such that localized fishers at least temporarily claim ownership rights to certain fish stocks. McCay concluded, however, that "most known cases of indigenous fisheries management . . . hinge upon the management of access to fishing *space* rather than levels of fishing effort" (p. 6). However, while we agree that folk management is often just that, we also see a proliferation of case studies showing that it is sometimes much more.

We conceive of folk management as an aspect of a people's fishing technology, construing *technology* broadly to include not only fishing gear and knowledge about its appropriate use, but also knowledge about how fishing effort should be organized and how a localized fishery should be managed. This fundamentally implies that a fishery is not merely a fish stock, but instead is the articulation through "technology" of a fishing community with certain living marine resources. In fact, there is no fishery in an aquatic environment unless there is human interaction with fish stocks, and it is this interaction that generates the ideational culture of folk management in fishing communities. Folk management, therefore, comprises many components of ideational culture, including ideas about what are the preferred and most appropriate types of fishing gear, how fishing activities should proceed, and how the fishery should be managed so that it is sustainable.

Few fisheries have escaped the resource-management problems of the modern world, and fewer still are managed under pristine and fully comprehensive systems of folk management. Indeed, knowledge associated with folk systems is being lost as many indigenous peoples face biological and cultural extinction (Bodley 1991). And as "progress" encroaches on folk systems, their management processes are inordinately influenced by externally imposed regimes. In such cases, therefore, much of what might be described as folk management is also a direct response, or reaction, to those externally imposed regimes. Regardless, we still feel it would behoove fishery managers to know more about even these reactionary types of folk management, which coexist alongside externally imposed regimes — often in spite of them — for these are still important signposts indicating how local fishing peoples feel the fisheries they work in should be managed.

On the other hand, although several new studies have recently come to light concerning folk management in fisheries of long ago — those describing Japan's coastal fisheries prior to the modern era, for example — we are aware that the applicability of these studies to managing modern fisheries must be considered very carefully, as management needs in premodern and modern situations may have few analogues. What is still of great potential value in these cases, nevertheless, is that they suggest a wide range of possibilities concerning what fishers may do, or want to do, to manage the natural resources upon which they depend.

One also should not assume that we feel folk management is inherently superior, and thus preferable, to all modern systems of management that are imposed by state authorities. There is no question that in many fisheries today, the local fishers are unable to manage without considerable outside help. Certainly the uncontrolled introduction of modern fishing technology in some fisheries has created a destabilizing impact on sustainable management, with the breakdown of formerly effective folk systems an important outcome. This is particularly true in fisheries that are experiencing resource depletion and high degrees of competition, or in which there is great diversity among the various fishers themselves in terms of such variables as their capital commitments, fishing gear, ethnicity, or socioeconomic status. In those situations, the "folk" have such widely divergent interests that, practically speaking, it may be impossible for them to collectively institute workable home-grown management regimes, at least not without considerable help and guidance from the outside (see Bailey 1988:119).

At this point some readers may question the utility for modern fisheries management of examining folk management, reminding us that, strictly speaking, there are almost no true "folk societies" left in the world. They may insist that the global economy, rapid transit, and rapid communications in particular have at least partially connected all former folk societies with the emerging global village, and that folk management as a subject would now best be left to fisheries historians. We wish to reorient this argument, however, by urging our readers to consider as "folk managers" any fishers who are linked to a renewable resource, regardless of their degree of cultural, technological, or socioeconomic sophistication. Thus, we are not particularly concerned with the often-presented dichotomy of small-scale and artisanal versus industrial fishers, but instead want to reveal and understand what works, or doesn't work, for "fisher folk" of practically any kind.

Some of our readers may also point out that the character of most fishing today, even that which is small- and medium-scale, is now inordinately influenced by the utilization of material technologies that have been adopted only fairly recently, and that, correspondingly, the influence of folk institutions has markedly declined. Thus, it might be argued, the new approaches to fishing that are necessitated by the use of these material technologies have muted the force and broken the continuity of any longstanding traditions of folk management that heretofore were effective means of localized fisheries management.

These are compelling arguments, and we agree that most "folk societies," at least as they have been conceptualized by social scientists (and especially anthropologists), have all but disappeared in the modern era. But these arguments all miss the main point for our purposes here, which is that a localized worldview, and locally developed assertions about how to best manage fisheries, still arise among fishing peoples at every level of technological sophistication, and that a better understanding of these processes will be important if fisheries management is to be more effective in the near future. Certainly many contemporary fishers are aware of the existence of peoples having very different cultures who live far beyond the regions in which they work and live, and many routinely communicate with relatives, friends, and others who live at great distances from them and who are not involved in fishing in any way. But otherwise, for them, where they live and work is still a localized, specific place, and quite often they perceive that they take their catches from a specific, bounded, marine ecosystem, which from their perspective has unique systemic attributes.

Their association with a specific place allows us to think of fishers as members of a special community type: the "natural resource community" (NRC) (Dyer, Gill, and Picou 1992; Dyer 1993). The natural resource community has been defined as "a population of individuals living within a bounded area whose primary cultural existence is based on the utilization of renewable natural resources" (Dyer, Gill, and Picou 1992:106). By being culturally integrated with the resources they utilize, the worldview of fishers is shaped and sustained by membership in NRCs. Moreover, fishers' appreciation of the sustainability of utilized natural resources creates a sense of environmental awareness that manifests itself as folk management.

Strong ties to NRCs are especially prevalent among the world's small- and medium-scale fishers, who constitute more than 90 percent of all the primary producers involved in fishing around the world (see Thomson 1980:3, Bailey 1988, Ben-Yami 1980, and Lawson 1984). Such fishers seldom range very far from their home ports, always have very definite ideas about the productivity of the marine ecosystems they exploit, and, it has been increasingly documented, always have very definite ideas about how the fisheries they work in should be managed.

All fisheries-management regimes must address two fundamental, and consequently separate, problems: first, how to conserve adequate stocks of marine resources, and second, how to fairly and equitably allocate that portion of these stocks so they can be harvested by fishers. Obviously, the first problem is mainly one for marine biologists. But the second one, which today is somewhat informed by formal economic theory, might be more richly informed by fishing peoples and social scientists who are knowledgeable about folk management.

Moreover, we should not be too hard on fisheries scientists and managers if they neglected folk management during the early development of fisheries-management science. Until fairly recently there was not much published information about fishing societies and folk management available to them. In fact, during the early development of fisheries-management science, the few social scientists who were interested in fishing peoples concerned themselves mainly with primitive fishing gear and exotic lifestyles, only rarely linking their studies to the concerns of emerging fisheries management. Indeed, it was not until the middle of the twentieth century, after the first crises in fisheries had launched the development of modern fisheries management, that the first truly comprehensive studies of fishing peoples having relevance to management concerns began to be published. Notable among these studies are Firth's *Malay Fishermen: Their Peasant Economy* (1946), Norbeck's *Takashima: A Japanese Fishing Village* (1954), and Fraser's (1960) *Rusembilan: A Malay Fishing Village in Southern Thailand.* From these landmark works, a new discipline, maritime anthropology, has emerged that is mainly concerned with fishing peoples and maritime adaptations, and to a lesser extent with early navigation, prehistoric maritime peoples, and several other miscellaneous topics. Because most maritime peoples are fishing peoples, it is understandable that this relatively new discipline has already become highly involved in debates concerning how best to manage the fisheries.

Studies explicitly concerned with folk management first appeared in the 1970s. Important pioneering works include two articles by Johannes (1977, 1978) and Klee's own chapter in his edited volume on traditional resource management (1980:245–281), with both authors describing folk management in Oceania. Also appearing in the 1970s was an article by Stiles (1972) and another by Andersen (1979), who both describe localized management in Canadian fisheries.

By now, however, there has been a virtual florescence of studies focusing on folk management. Notable works include Acheson (1987); Berkes (1986, 1987, 1989a, 1989b); Dahl (1988); Davis (1985); Durrenberger and Pálsson (1987); Jentoft (1985); the several contributions by Berkes, Carrier, and Hames in McCay and Acheson (1987); McGoodwin (1984, 1990:112–142); Morauta, Pernetta, and Heaney (1982); and Pinkerton (1989). Noteworthy book-length works include Cordell, *A Sea of Small Boats* (1989), Ruddle and Akimichi, *Maritime Institutions in the Western Pacific* (1984), and Ruddle and Johannes, *The Traditional Knowledge and Management of Coastal Systems in Asia and the Pacific* (1985).

Regretfully, this new literature has not made much impact in fisheries-management circles. One reason is that these works are rather new and not yet widely known. Another may be that from a purely scientific perspective they seem rather disparate, which makes their lessons hard to distill. We share this concern and have therefore urged each of our contributors to address certain core issues in fisheries management, as well as to conclude their contributions with a series of summarizing statements under the subheading "Lessons for Modern Fisheries Management." This, we hope, will help bring about some uniformity among the various contributions, as well as provide concise summaries for readers who are especially pressed for time.

Certainly a perduring core issue in fisheries management has been common-property theory. As Ruddle and Akimichi say of fishers, "their uncertain, weak or contested tenurial status is one of the principal difficulties encountered by small-scale fishermen" (1984:1). This problem seemed engraved in stone some twenty-five years ago, with the publication of Garrett Hardin's landmark article, "The Tragedy of the Commons" (1968). Indeed, because most fisheries are instituted as common-property resources, localized fishers often find it extremely difficult to lay claim to the resources they go after. This makes the future availability of such resources problematic for them, because the more successful they are, the more likely it is that others will enter the fishery to compete with them. Thus, Hardin believed, it was only "rational" for them to take all they could today, hastening the depletion of the fishery, because there might be no resources for them to harvest tomorrow — a scenario that has been acted out many times, with tragic consequences, in many contemporary fisheries.

However, we feel certain that a fuller understanding of folk management holds great promise for solving, reconceptualizing, or at least mitigating the common-property problem in fisheries, mostly because many alternatives for management exist in fisheries that are otherwise instituted as common-property resources. Recent work in maritime anthropology, as well as in several other disciplines concerned with resource management, shows that the "tragedy of the commons" theory, at least as it is customarily phrased, has many shortcomings. Now, a growing body of case studies shows how localized fishing peoples have developed fisheries-management strategies that avoid the development of the tragedy in fisheries that are instituted as common-property resources.

Actually, what most often seems to bring about the breakdown of localized systems of control of common-property resources is the articulation of a fishery with modern marketing systems while permitting open access. Thus, even when fishers who exploit a common-property fishery are highly articulated with modern seafood markets, they may not be prompted to overexploit the fishery if they can limit access to it, particularly when by limiting access they can ensure higher sustained prices for their catches.

Nevertheless, many students of the fisheries see modernization and the mentalities that go with it as mainly responsible for the breakdown of socially imposed systems of self-restraint among fishers. Modern mentalities are different mentalities, they insist, which stress productivity, competitiveness, and individualism over the needs of society as a whole. Such assertions are clearly contradicted, however, in the documentation of robust systems of folk management existing among otherwise modern and contemporary fishing peoples. Indeed, most of the case studies in this book concern fully modern fishers, contemporary "folk," who live in today's modern world.

Knowing more about folk management will not be a panacea for all that ails fisheries management these days, but it does seem a promising next step. Nowadays, quite unlike when fisheries-management science was first born, there is abundant social-scientific information about fishing peoples. By learning about the managerial strategies localized fishers prefer and employ, while encouraging their participation at every stage in the development of management policies, a modern manager goes with the local cultural grain. A natural process is capitalized upon, rather than being opposed or confounded by the imposition of an exogenous one. Management policies are encouraged to grow from the bottom up, beginning with an accumulation of inputs from the people who will be most impacted — fishers themselves.

Ideally, fisheries management should proceed as a collaborative effort between professional managers and fishers. Fisheries managers who have an intimate understanding of the people they are charged with managing, who know what management strategies these people are likely to invent, and what strategies they typically prefer, will have a far better chance of getting cooperation than will managers who keep their distance and remain focused mainly on conservation of resources. Likewise, fishers who are made aware of the culture of fisheries managers, who are given examples of successfully managed fisheries, and who are encouraged to actively engage managers in a cooperative dialogue can empower themselves to achieve greater control and sustainability of the natural resources they exploit.

The case studies in this book cover a wide range of issues in folk management in the world's fisheries. Pomeroy's discussion of Lake Chapala fisheries in Mexico (Chapter 1) provides an analysis of cultural and managerial obstacles to the institutional development of sustainable folk-managed fisheries. She demonstrates that lack of support or interference with folk systems by managers can impair their development or destroy their viability. This state of affairs is perpetuated by a system of favoritism and corruption among state managers.

McGoodwin's contribution (Chapter 2) portrays a similar scenario for an artisanal shrimp fishery on Mexico's Pacific coast. Here, overcapitalization, increasing competition, and mismanagement led to the destruction of a folk system based on mutual respect of fishing territories and rules of first-come, first-served.

In their study of the oyster fishery of the Gulf of Mexico (Chapter 3), Dyer and Leard compare open and closed natural resource communities having varying levels of folk management. In the closed NRCs of Florida and Louisiana, folk management and state-supported co-management have succeeded in providing for sustainable harvests and economic stability. On the other hand, open NRCs with weak systems of folk management in Alabama and Mississippi are shown to be much less viable, and subject to economic instability and greater variation in productivity in oyster harvests.

Ward and Weeks (Chapter 4) explore the basis of conflict between managers and users in the oyster fishery of Texas. Conflicts arise from differences in training and worldview, resulting in a proprietary attitude toward the resource by managers that inhibits the integration of users' folk knowledge into management policy. In Chapter 5, Stoffle et al. describe a viable case of folk management in a reef fishery in the Dominican Republic. The anthropologists' fieldwork helped validate a case for local control of the reef. Their case was strengthened by satellite imaging techniques showing the superior health of folk-managed reefs in comparison to surrounding unmanaged reefs.

In Chapter 6, Anderson provides an analysis of the importance of religious sanctions as a means of conserving local fishery resources among indigenous peoples along the northwest coast of North America, as well as among Cantonese fishers of Hong Kong. Sanctions are linked to an intimate understanding of the biology of exploited species, or folk knowledge. In Chapter 7, Ruddle gives a rich overview of the significance of folk knowledge in the management of marine and estuarine fisheries of the South Pacific. He points out the significance of such knowledge in fisheries where limited resources for generating "scientific" data can be augmented by use of available folk knowledge of local fisheries.

Gill (Chapter 8) describes the impact of the *Exxon Valdez* oil spill on co-managed salmon fisheries in Prince William Sound, Alaska. Negative social, psychological, and ecological impacts of this technological disaster on local communities resulted in disruptions to the cooperative basis of the co-management system. This case illuminates the threat posed by pollution to the viability of folk- managed fisheries and provides some potential solutions.

Palmer's contribution on the lobster fisheries of Maine and Newfoundland (Chapter 9) provides evidence for the inadequacy of folk systems of control as a panacea for sustainability. Here, state intervention was necessary to save the fishery from collapse and was regarded by the local lobster fishers as an acceptable action.

Felt (Chapter 10) also deals with folk perceptions of resource control. He demonstrates how the construction of social knowledge of resource viability among the salmon fishers of Atlantic Canada can be linked to variations in the comparative health of ecologically distinct fish stocks. Jentoft and Mikalsen (Chapter 11) provide an example of how regional interest groups can undermine localized folk-management efforts in the fjord fisheries of Norway. They point out that political machinations are one of the most difficult aspects to deal with in management and can subvert the most well-intended, user-oriented policies.

Pinkerton's "Summary and Conclusions" integrates various of the contributions into useful ordinal models that bring out the major problems and issues of folk, state, and co-management.

This book takes a step toward defining and conceptualizing folk management in the world's fisheries. We hope that it finds use not only as a scholarly contribution to maritime anthropology, but also as a jumping-off point for managers, social scientists, fishing organizations, and fisher folk around the world as they seek viable solutions to the shared problems of contemporary fisheries management.

References

Acheson, James M.
    1987   The Lobster Fiefs Revisited: Economic and Ecological Effects of Terri-
           toriality in Maine Lobster Fishing. Pp. 37–65 in Bonnie J. McCay and
           James M. Acheson, eds., The Question of the Commons: The Culture
           and Ecology of Communal Resources. Tucson: University of Arizona
           Press.

Andersen, Raoul
    1979   Public and Private Access Management in Newfoundland Fishing. In R.
           Andersen, ed., North Atlantic Maritime Cultures. The Hague, Nether-
           lands: Mouton.

Bailey, Conner
    1988   Optimal Development of Third World Fisheries. Pp. 105–128 in Michael
           A. Morris, ed., North-South Perspectives on Marine Policy. Westview

Special Studies in Ocean Science and Policy. Boulder, Colo.: Westview Press.

Ben-Yami, Menachen
  1980 Community Fisheries Centres and the Transfer of Technology to Small-Scale Fisheries. Pp. 936–948 in Proceedings of the 19th Session of the Indo-Pacific Fisheries Council, Kyoto, Japan, May 21–30.

Berkes, Fikret
  1986 Local Level Management and the Commons Problem. Marine Policy 10:215–229.
  1987 Common-Property Resource Management and Cree Indian Fisheries in Subarctic Canada. Pp. 66–91 in Bonnie J. McCay and James M. Acheson, eds., The Question of the Commons: The Culture and Ecology of Communal Resources. Tucson: University of Arizona Press.
  1989a Local-Level Resource Management Studies and Programs: The Great Lakes Region and Ontario. Pp. 95–118 in F. G. Cohen and A. J. Hanson, eds., Local-Level Resource Management in Canada: An Inventory of Studies and Programs. Ottawa, Canada: MAB/UNESCO.
  1989b Ed. Common Property Resources: Ecology and Community-Based Sustainable Development. New York: Columbia University Press.

Bodley, John H.
  1991 Victims of Progress. 3rd ed. Palo Alto, Calif.: Mayfield Publishing.

Carrier, James G.
  1987 Marine Tenure and Conservation in Papua New Guinea: Problems in Interpretation. Pp. 142–167 in Bonnie J. McCay and James M. Acheson, eds., The Question of the Commons: The Cultural Ecology of Communal Resources. Tucson: University of Arizona Press.

Cordell, John, ed.
  1989 A Sea of Small Boats. Cultural Survival Report 26. Cambridge, Mass.: Cultural Survival.

Dahl, C.
  1988 Traditional Marine Tenure: A Basis for Artisanal Fisheries Management. Marine Policy: 40–48.

Davis, A.
  1985 Property Rights and Access Management in the Small Boat Fishery: A Case Study From South Western Nova Scotia. In C. Lamson and A. Hanson, eds., Atlantic Fisheries and Coastal Communities: Fisheries Decision-Making Case Studies. Halifax, Nova Scotia: Dalhousie University.

Durrenberger, E. Paul, and Gísli Pálsson
  1987 Ownership at Sea: Fishing Territories and Access to Sea Resources. American Ethnologist 14(3):508–522.

Dyer, Christopher L.
    1993 Tradition Loss as Secondary Disaster: Long-Term Cultural Impacts of
         the *Exxon Valdez* Oil Spill. Sociological Spectrum 13:65–88.

Dyer, Christopher L., Duane A. Gill, and J. Steven Picou
    1992 Social Disruption and the *Valdez* Oil Spill: Alaskan Natives in a Natural
         Resource Community. Sociological Spectrum 12:105–126.

Firth, Raymond
    1946 Malay Fishermen: Their Peasant Economy. London: Routledge and
         Kegan Paul.

Fraser, Thomas M., Jr.
    1960 Rusembilan: A Malay Fishing Village in Southern Thailand. Ithaca, N.Y.:
         Cornell University Press.
    1966 Fishermen of Southern Thailand: The Malay Villagers. New York: Holt,
         Rhinehart, and Winston.

Hames, Raymond
    1987 Game Conservation or Efficient Hunting? Pp. 92–107 in Bonnie J.
         McCay and James M. Acheson, eds., The Question of the Commons:
         The Culture and Ecology of Communal Resources. Tucson: University
         of Arizona Press.

Hardin, Garrett
    1968 The Tragedy of the Commons. Science 162:1234–1248.

Jentoft, Svein
    1985 Models of Fishery Development: The Cooperative Approach. Marine
         Policy 9:322–331.

Johannes, R. E.
    1977 Traditional Law of the Sea in Micronesia. Micronesia 13(2):121–127.
    1978 Traditional Marine Conservation Methods in Oceania, and Their De-
         mise. Annual Review of Ecology and Systematics 9:349–364.

Klee, Gary A., ed.
    1980 World Systems of Traditional Resource Management. New York: John
         Wiley and Sons.

Lawson, R.
    1984 Economics of Fisheries Development. New York: Praeger.

McCay, Bonnie J.
    1981 Development Issues in Fisheries as Agrarian Systems. Culture and
         Agriculture 11(May). Bulletin of the Anthropological Study Group on
         Agrarian Systems, no. 11. Urbana-Champaign: University of Illinois.

McCay, Bonnie J., and James M. Acheson, eds.
    1987 The Question of the Commons: The Culture and Ecology of Communal
         Resources. Tucson: University of Arizona Press.

McGoodwin, James R.
    1984   Some Examples of Self-Regulatory Mechanisms in Unmanaged Fisher-
            ies. Pp. 41–61 in Food and Agriculture Organization, Expert Consul-
            tation on the Regulation of Fishing Effort (Mortality). FAO Fisheries
            Report No. 289, Supplement 2.
    1990   Crisis in the World's Fisheries: People, Problems, and Policies. Stanford,
            Calif.: Stanford Univerity Press.

Morauta, L., J. Pernetta, and W. Heaney, eds.
    1982   Traditional Conservation in Papua New Guinea: Implications for Today.
            Monograph no. 16. Boroko, Papua New Guinea: Institute of Applied
            Social and Economic Research.

Norbeck, Edward
    1954   Takashima: A Japanese Fishing Village. Salt Lake City: University of
            Utah Press.

Pinkerton, Evelyn W., ed.
    1989   Co-Operative Management of Local Fisheries: New Directions for Im-
            proving Management and Community Development. Vancouver: Uni-
            versity of British Columbia Press.

Ruddle, Kenneth, and Tomoya Akimichi, eds.
    1984   Maritime Institutions in the Western Pacific. Senri Ethnological Studies
            17. Osaka, Japan: National Museum of Ethnology.

Ruddle, Kenneth, and R. E. Johannes, eds.
    1985   The Traditional Knowledge and Management of Coastal Systems in Asia
            and the Pacific. Jakarta, Indonesia: Regional Office for Science and
            Technology for Southeast Asia, UNESCO.

Stiles, R. G.
    1972   Fishermen, Wives, and Radios: Aspects of Communication in a New-
            foundland Fishing Community. Pp. 35–60 in R. Andersen and C. Wadel,
            eds., North Atlantic Fishermen: Anthropological Essays on Modern
            Fishing. Toronto, Ontario: University of Toronto Press.

Thomson, David B.
    1980   Conflict Within the Fishing Industry. ICLARM Newsletter 3(3):3–4.

1

Obstacles to Institutional Development in the Fishery of Lake Chapala, Mexico

Caroline Pomeroy

Fisheries possess characteristics that make them appropriate for common pool use (Oakerson 1986) and, by the same token, susceptible to destruction in a "tragedy of the commons" (Hardin 1968). Cooperation among resource users to avert the tragedy, however, occurs as individual, voluntary modification of behavior or as participation in institutional development, that is, in the creation and maintenance of rules for coordinating resource use. Field researchers have documented such institutions (see McCay and Acheson 1987, McGoodwin 1990, and Pinkerton 1989), which resource users have either created among themselves or have appropriated from external sources and adapted to local conditions. Both are forms of folk management, whereby resource users can participate in the coordination of resource use, ideally to insure its sustainability. Although some of these systems are self-contained, others are integrated with external management.

Under certain conditions, local common pool resource (CPR) management institutions offer low-cost alternatives or comple-ments to expensive, and often inappropriate, external (govern-ment) management. Yet, the success of local institutional development in general, and folk management in particular, in a CPR setting is affected by a complex set of environmental, social,

First presented at the Third Conference of the International Association for the Study of Common Property, Washington, DC, September 1992. Author's current affiliation: Workshop in Political Theory and Policy Analysis, Indiana University, Bloomington, IN 47405–3186.

and political factors. These factors constitute the conditions under which local resource-management institutions operate. Researchers have sought to specify conditions that are favorable to local institutional development and survival. Ostrom (1990), for example, has formulated a set of hypothesized conditions, termed design principles, for the development of successful institutions for local resource management.[1] Institutional analysis guided by these design principles involves the exploration of situational factors that affect conditions and outcomes in the CPR setting.

These design principles helped guide this study of the small-scale commercial fishery of Lake Chapala, Mexico (see Figure 1.1). The purpose of the study was to explore the relevance of selected individual and organizational factors to cooperation in CPR use. From October 1991 through April 1992, data were collected using participant observation, personal interviews, documentary and archival research, and a survey interview. Informants included fishers, buyers, local and state Fisheries Secretariat (Secretaría de Pesca, known as PESCA) authorities, and others familiar with the Chapala fishery. The survey interview was conducted with a random sample of fishers from one large union and with all members of two smaller organizations to limit overrepresentation of the considerably larger group; 130 survey interviews were completed.

The two most important design principles identified in this study are clear boundary definition and authorities' recognition of resource users' rights to organize. To address the relevance of boundary definition and recognition by authorities to effective resource management at Lake Chapala, I begin with a brief description of the research and its setting. Next, I describe the boundaries of the fishery as articulated in law and government policy and in policy espoused by the fishers' organizations studied. Then, I discuss examples of inconsistencies in authorities' recognition of these various boundary definitions and the effects these inconsistencies have on fishers. Finally, I summarize the combined effects of unclear boundary definition and incomplete recognition of

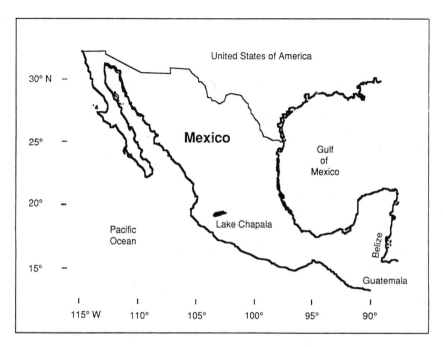

Figure 1.1. Mexico and the location of Lake Chapala.

fishers' efforts to coordinate resource use on their ability to resolve problems associated with that use.

The Setting

Lake Chapala, located forty-eight kilometers southeast of Guadalajara, Jalisco, is the third-largest lake in Latin America. The lake is the focal point of the five-state Lerma-Chapala-Santiago watershed and itself covers 1,100 square kilometers in two states, Jalisco and Michoacán. In addition to providing fishing opportunities at the lake, the watershed supports a variety of uses, including irrigation agriculture, municipal water supply, and tourism. These uses, coupled with a decade of dry weather, were responsible for substantial losses of water quality and quantity, which led to spatial crowding in the fishery and apparent shortages and contamination

of fish stocks. Although heavy rains in 1991 and the implementation of various infrastructural improvements (e.g., sewage treatment plants) have limited further deterioration of water quality and
allowed the lake level to rise, these problems and others associated
with use of the CPR persist.

My study focuses on three organizations and their members:
Unión de Pescadores del Lago de Chapala ("the Chapala union"),
Sociedad Cooperativa de Producción Pesquera Pescadores de Chapala ("the cooperative"), and Unión de Pescadores San Pedro
Tesistán ("the San Pedro union"). The first two organizations have
approximately 180 and 30 members, respectively. Both are based
in Chapala, the municipality's center and the site of the regional
PESCA office. The San Pedro union has 20 members and is based
in San Pedro Tesistán, a small, south-shore lake community in the
municipality of Jocotepec (see Figure 1.2). All three organizations
fall under the jurisdiction of the regional PESCA office.[2]

The lake's small-scale commercial fishery involves an estimated 2,500 fishers, two-thirds of whom are registered in fifty-nine
unions and eight cooperatives (PESCA 1990a); the remainder com-
prise unregistered *pescadores libres,* or "free" fishers.[3] Fishers use
stationary gear, including nets, traps, and longlines, to catch tilapia,
carp, and catfish. They travel to and from gear sites in wooden or
fiberglass skiffs, some of which are equipped with outboard motors.
Mobile gear includes the *mangueadora* (haul net or seine) and the
atarraya (cast net), both used to catch *charal,* a local commercial
fish.[4] The mangueadora is fished by two people from a long, narrow
boat; the atarraya may be fished from a boat, from public shoreline,
or in individually concessioned federal parcels of land, known as
ranchos charaleros, or charal ranches.[5]

Most Chapala and San Pedro fishers sell their catch to buyers
who then sell the fish live, eviscerated, or as filets. Although there
are six stalls in Chapala's central market that sell lake fish, much
of the catch is sold on the street; a small fraction of it also is sold in
Guadalajara, the nearest major market center. In San Pedro, where
there is no market, fishers sell their catch to one of two buyers,

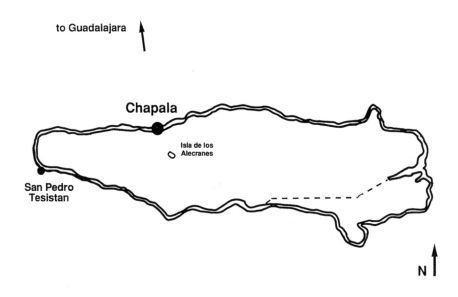

Figure 1.2. Lake Chapala.

both of whom are union members.[6] Charal are fished by Chapala fishers, but not by those in San Pedro. Charal fishers sell to one of three processors (two in Chapala, one in Petatán, Michoacán). Fishers dry some of the catch before selling it to buyers, but most is sold fresh for processing and local sale by street vendors or for bulk delivery to Monterrey and Mexico City.

Clearly Defined Boundaries

Clear definition of resource boundaries and who may and may not use the resource constitute the first step in organizing for collective action (Ostrom 1990). Boundary definition makes exclusion of potential resource users possible and helps assure resource users that they will be able to continue to derive benefits from the resource. Boundaries create a sense of control and efficacy (Edney

1981). Users' efforts to coordinate resource use are encouraged by a sense of control over investments they make in caring for the resource or otherwise coordinating its use. Without boundaries, a high demand for the resource increases the possibility of its destruction, because users adopt a 100 percent discount rate (Ostrom 1990). There is no assurance or guarantee that the resource will be there tomorrow if one forgoes taking it today.

Boundaries of the Lake Chapala fishery are variously prescribed by law and policy. These prescriptions are then interpreted and applied by authorities and fishers' organizations. The variety of prescriptions and interpretations means that boundaries are, in fact, *not* clearly defined, and this adds to the uncertainty associated with the use of a renewable resource.

Defining boundaries for Lake Chapala's fisheries is complex, as the fishery resource includes mainland and island shoreline where some fishing occurs, as well as open water and the fish targeted for capture.[7] Each of these elements is relevant to boundary definition. Legally defined, the fishery resource is not common property, but state property, to be managed primarily by the Fisheries Secretariat. According to federal law, inland waters, unless they lie within private property, are federal property (PESCA 1990b). Fish, likewise, are federal property, yet become private property upon capture by legitimate appropriators. Because fish are mobile, it is not possible to define areas within the lake for certain stocks or species, thereby making resource boundary definition difficult. Any adult Mexican citizens who have registered themselves and their fishing equipment with the National Fishing Registry and who have obtained permits and concessions required by federal, state, and local fishery authorities may fish commercially (PESCA 1987).[8]

The territorially based *rancho de charal* (charal, or whitefish, ranch) fishery entails legally specified boundaries tied to the concession and use of federal land. The ranchos fall under the jurisdiction of the Secretariat of Agriculture and Water Resources (Secretaría de Agricultura y Recursos Hidráulicos, or SARH). To insure exclusive right to a rancho, a fisher obtains a concession to a plot of shoreline land and its submerged extension from SARH through the National Water Commission (Comision Nacional de Aquas, or CNA). The concession is secured for a fee and by contract, which presently guarantees the holder a one-year exclusive right of usufruct (SARH 1989). Fishers value this exclusive right because they make improvements to the area to attract spawning charal, and they feel exclusively entitled to the benefits to be derived from this effort. As Ostrom (1990) notes, without defining boundaries and closing the resource to outsiders, local users face the risk that the benefits produced by their efforts will be reaped by others who have not contributed.

Fisheries policy, established primarily by the state PESCA delegation, provides a second set of resource and user boundary definitions.[9] Policy is the result of authorities' interpretations of the law for application to their jurisdiction. Furthermore, policy and its enactment reflect the extent of authorities' recognition of fishers' rights to devise their own institutions. Given sufficient and accurate information, policy, its interpretation, and its implementation can result in well-fitting rules and enforcement that build assurance among resource users. In the Chapala case, policy espoused by the state office in Guadalajara provides one interpretation of fishers' rights; that voiced by the Chapala office provides another. There is no policy coordination between the three regional PESCA offices (two in Jalisco, one in Michoacán) that share jurisdiction over lake fisheries.

PESCA policy upholds law designating the lake and its submerged lands as federal territory, freely accessible to any Mexican citizen with the required permits. At the same time, the local PESCA chief and staff acknowledge the tacit territorial division between Jalisco and Michoacán. That territorial distinction means that policy formulation and application by the local office and its superiors is confined to Jalisco's territory and its fishers. A more vague resource boundary exists between the jurisdictions of the two regional offices within the state. Within the Chapala office's region, which covers about 60 percent of the lake, local staff members have condoned and at times promoted the extension of community boundaries to fishing grounds. They encourage fishers to ask permission from another community's fishers if they want to fish within 500 meters of that community's shoreline. The policy has been backed, if weakly, by local PESCA officers' occasional visits to communities where fishers complain of intruders. The suggestion that fishers ask permission if they want to fish in non–home communities has come to be taken by some to be a rule, if not encoded law. Many fishers speak of a group's or community's zones, where resident fishers have preference in setting their gear. The delineation of such zones creates both social and physical conditions that assure fishers of their exclusive right, and it promotes coordination of resource use among them.

Many fishers view the establishment of community zones as a way to reduce conflict by limiting the number of and differences among fishers, thus facilitating their efforts to coordinate resource use. These effects of small group size and homogeneity are noted for their association with increased cooperation in social dilemmas (Edney 1981). By excluding outsiders, insiders are more likely to develop a sense of interdependency within the community, which can make widespread noncooperation among them implausible and enhance assurance that others within the group also will cooperate, or abide by the agreed-upon rules (Runge 1986). The establishment of boundaries facilitates monitoring and enforcement by the group of users by limiting the size of the resource use area. The visibility of a nearby, well-defined area enables fishers to monitor resource use and conditions within their bounds, thereby reducing some of the uncertainty associated with CPR use (Cass and Edney 1978; Jorgensen and Papciak 1981).

PESCA's policy defining who may and may not fish commercially (i.e., sell fish) departs from the legal definition and is ambiguous and inconsistently applied. In 1984, PESCA began to pursue the policy of requiring fishers to form unions and cooperatives. This policy was adopted as an expedient for management (Paré 1989).

It also allowed PESCA to break the power base of the Chapala union, which had been established in 1960 by a local *cacique* (political boss).[10] The rationale offered to fishers for forming more localized organizations was that attending meetings would be easier and less costly. PESCA created an additional incentive by offering aid (e.g., gear discounts, easier access to loans) to newly formed groups to replace what had long been supplied by the president of the original Chapala union. It was hoped and assumed that fishers would use the organizations as a mechanism for coordinating resource use, including commercialization, and for achieving social objectives such as improved well-being of fishers and their families. Permits, which had been required since 1971, continued to be issued to individuals until 1989, when the state office shifted to a policy of granting permits only to organizations, thus requiring commercial fishers to join a group to obtain permits and fish legally.

Local PESCA policy does not explicitly limit entry to the fishery. Rather, the Chapala PESCA chief has deferred the decision to admit new fishers to fishing organizations. PESCA's permitting policy notwithstanding, an individual permit can still be obtained by going directly to the Guadalajara office, but this procedure is costly and time-consuming. In addition to spending time and money to go to Guadalajara, one must pay the 114,000 pesos (about $47 U.S.) fee per permit. By joining a group, one saves the former costs because a group representative obtains the required permits locally. The permit cost is significantly diminished because it is divided among group members.

Yet another set of boundary definitions is found among fishers and in group policy on admitting and excluding members. Most fishers interviewed share the opinion that the lake is "free," meaning one can fish where one pleases, provided one respects others' gear. Anyone who has obtained required permits and carries a *credencial,* or identification card, may use the resource, but all others are excluded. As most fishers understand the rules, one can obtain these things only by joining a group. The resource is perceived as the common property of all fishers who are members of a union or cooperative. Group membership entails the costs of monthly dues, permit fees, and meeting attendance.[11] Benefits include the legal protection of coverage by a group fishing permit and assistance if one is in economic or legal trouble. Organized fishers who regularly attend meetings criticize absent members for being *de pura convenencia*, or free riders who avail themselves of these benefits without paying the costs.

Boundary definition is more explicit in San Pedro, where fishers say that as union members, they should have the exclusive right to fish the zone adjacent to their community. This idea derives in part from their past association with a coalition of fishers' organizations. The coalition espoused group- or community-specific territories and presented the idea to PESCA as a means of rationalizing the fishery (Paré 1989). San Pedro fishers report intrusions by outsiders from distant communities and complain of the unfairness of local free fishers, who lack membership in a union or cooperative, lack a permit, or both. Union fishers feel they are abiding by the rules and paying the costs to fish, and these others are not. Free fishers are free riders, who secure benefits from the fishery (e.g., access to the resource and income from its sale) without contributing to the costs (e.g., permit fees, group participation).

The PESCA chief's deferral of decisions to admit new fishers to the unions and cooperatives has given the groups an opportunity, albeit limited, to define who may or may not use the resource. Each group has its own criteria for admitting and excluding (both by denying entry and expelling) members. Stealing others' fish or gear and other disruptive behavior are grounds for expulsion.

In San Pedro, anyone who demonstrates an interest in the fishery and the group is likely to be admitted. The prospective member must agree to attend meetings and pay a share of the permit and other group dues and may be called on from time to time to contribute labor or money to a group or community project. Although the union did not receive requests for admission during the study, a core of active members was trying to redefine group boundaries by excluding noncomplying members. Noncomplying members are those who have (1) moved to the United States without giving notice to the group or paying fees to cover their absence or (2) stopped attending meetings and remain in the community, some of them continuing to fish. To exclude members, the union gave the local PESCA officer a list of active members, including two who had given notice and paid fees before going to the United States for several months. Union members also provided the PESCA officer with the names of noncomplying members, as well as others who they say are free fishers, that is, fishers who are not affiliated with a group and lack permits.[12] Fishers removed from the list no longer are covered by a permit and would be fishing illegally.

As in San Pedro, the Chapala cooperative is trying to redefine group boundaries by excluding noncomplying members. Admitting and excluding members from the cooperative, however, is more complicated, due to the Federal Cooperative Law and regulations (SARH 1938a, 1938b), which establish explicit criteria and relatively complex procedures for admission and exclusion of members. Changes in group membership must be submitted to the Secretariat of Labor and Social Provision (Secretaria del Trabajo y Prevension Social) in Guadalajara and Mexico City and to the state Federation of Fishing Cooperatives ("the Federation"), a nongovernment coalition, for approval. This procedure occurs independent of PESCA, although the cooperative is supposed to notify the local PESCA office for purposes of permitting and keeping track of fishers. In observed monthly meetings, a core of active members tried to redefine group boundaries by excluding members who had stopped participating in group activities.[13] Daunted by the procedure outlined by cooperative regulations, they sought help from the Federation. Although the cooperative leadership sent the Federation a revised list of members, and later pursued the matter at its state headquarters in Barra de Navidad, no progress had been made by the end of the study.

The Chapala union's policy is to admit sons of fishers, virtually without limit; to provisionally accept other fishers; and to reject those who intend to enter the fishery for only a brief period.[14] This policy departs notably from that of the previous union leadership, which actively sought new members. New members admitted during the study included two fishers who had had problems with the leadership of a nearby union, and four fishers from another lake who are related to several union members. Entry was denied to a man who wanted to fish during Lent but had no prior experience or intention of fishing at other times. Members explained that by gaining temporary membership, he would derive the short-term benefits of the brief high market season and the legal protection afforded by group membership without paying the full costs of group participation. Like the other two organizations studied, the Chapala union has begun to exclude members for failing to comply with group rules (e.g., attending monthly meetings, paying dues). At the November 1991 union meeting, the treasurer read a list of members to be excluded for their failure to contribute to an annual religious festival in which fishers play a large part. Over the next several months, fishers were given the chance to pay the fee. The list was finalized in March 1992 to be delivered to PESCA, which could then enforce the permit requirement and effectively exclude some resource users.[15]

Minimal Recognition of Rights to Organize

The problems associated with the lack of clearly defined boundaries are exacerbated by inconsistencies in authorities' recognition of fishers' rights to organize. According to Ostrom, minimal recognition means that "the rights of appropriators to devise their own institutions are not challenged by external government authorities" (1990:101). Resource users may make their own rules or modify externally designed rules to best fit local conditions. To enable enforcement of locally devised or adapted rules, they must be recognized by government resource-management authorities as legitimate. If government limits fishers' enforcement authority, it must then provide a consistent and workable alternative. Otherwise, the bounded system will become open access, and assurance problems will persist.

At Lake Chapala, inconsistencies in authorities' recognition of fishers' rights to organize have destabilized boundary definitions and further inhibited the emergence and maintenance of local institutions for CPR use. Some examples are (1) the legal recognition of the rancho system, (2) the definition of fishing zones for each organization, (3) support of groups' efforts to redefine user boundaries, and (4) the recognition of a coalition of fishers' organizations.

Ranchos charaleros were first developed in the late 1800s on federal lands as an extension of lakeshore cultivation. Following the Mexican Revolution (1910–1924), the government required individuals to obtain concessions to these federal lands. The rancho system was legally recognized in the 1930 Fishery Code, which specified the exclusive right of concession holders to fish these areas. The code also exempted rancho fishers from the charal closure, based on the argument that they enhanced charal reproduction by creating and maintaining spawning habitat and by scaring away predators (Departamento Forestal de Caza y Pesca 1939). In the revised 1985 code, however, recognition of these exclusive rights is absent, as is any reference to freshwater fisheries. The omission is attributed to the simplification of the fishery code over the intervening years and to the relative economic unimportance of inland compared to coastal fisheries.

The local PESCA chief points to the discrepancy between the 1930 and 1985 codes and has used it in his efforts to gain legal recognition for the rancho system. He makes two arguments in favor of fully legalizing the ranchos. First, it makes more administrative and practical sense to legitimize and rationalize this fishery than to fight it (which would be costly to both PESCA and fishers). Recognition of the rancho system would enable PESCA to exercise some control over it as well (e.g., define boundaries, collect revenues). Second, the ranchos are an extensive aquaculture system in which fishers create spawning habitat and drive predators away while tending and fishing their ranchos. By enhancing production in their ranchos, fishers assume some of PESCA's resource-management responsibilities. A third argument in favor of the rancho system relates to its history. As a local institution that developed informally among resource users, it entails a set of rules and mechanisms to promote reasonable use, which have been adapted to changing conditions in the fishery. This system can be and often is used without recourse to authorities and thus is more cost-effective than direct government management. The value of such territorial-based systems has been documented in other fisheries (e.g., Panayotou 1983, LeVieil 1987).

At Chapala, however, recent changes in resource conditions, coupled with social and political factors, have increased pressure on the system, such that its ability to adapt may not be sufficient to avert problems associated with the use of a scarce shared resource. Under current water resource and fishery laws, the ranchos are not legalized, nor are fishers' rights to work them during the charal closure. The contract for a federal zone concession was designed for farmers who wanted to lease fertile lakebed land, and thus prohibits changes to the landscape other than seasonal ones. Modifications fishers make to enhance charal production and reproduction (e.g., constructing jetties, clearing lake bottom of debris, planting shade trees) are perceived as more permanent, intended to create good, enduring charal habitat. Although state and local PESCA authorities have sought revisions of the contract, no adjustments have been made to recognize fishers' customary right to make these modifications. The Secretariat of Agriculture and Water Resources (SARH) and the Secretariat of Urban Development and Ecology (Secretaria de Desarollo Urbano y Ecologia, or SEDUE) are empowered to revoke the concession for violation of contract terms (SARH 1989). Also, to increase its revenue from and control over federal zone land tenure, SARH canceled all ninety-nine-year concessions and replaced them with one-year contracts in 1989. Federal PESCA authorities do not acknowledge the rancho system's value as an aquaculture project, nor the customary right of rancho holders to fish during the charal closure. Thus the exclusiveness of fishers' rights to their ranchos and to the benefits they derive from their investments in tending them are neither clear nor stable. SARH's refusal to modify the contract to recognize the legitimacy of the rancho system and federal PESCA authorities' lack of acknowledgement are examples of obstructive policy undercutting a local institution (Lawry 1990).

The only assurance fishers have is a written agreement between Chapala union fishers and state and local PESCA authorities that recognizes fishers' rights to fish their ranchos according to custom as long as they continue to tend the ranchos. The value of the rancho system as an aquaculture experiment is cited in the accord as the principal justification for the exemption (Ortiz 1989). As official recognition of the rancho system, the accord serves as a deterrent, albeit a weak one, to SARH's or SEDUE's taking action against fishers for contract violations.

Although the accord extends tacitly to other organizations whose members operate ranchos elsewhere in the Chapala region, providing some assurance of their exclusive rights, it also has had a negative effect on those rancho holders who are cooperative members. The union-PESCA accord was made when a farmer bid for a long-term concession of Isla de los Alacranes. At the time, there were only a few ranchos on the island, held by fishers from both south- and north-shore communities. To strengthen rancho fishers' claim to the island, PESCA helped secure the concession from SARH in the name of the Chapala union, rather than in the name of individuals. Union fishers then gained preference to future island concessions over fishers from other groups because they were named explicitly in the accord. Rancho fishers from other communities were pressured to join the Chapala union or give up their ranchos on the island. Fishers from the cooperative refused to do either, and finally an accord was established between the two groups that allotted the *islote,* a small sandbar adjacent to Alacranes, to the cooperative. Cooperative members feel cheated by the arrangement because islote ranchos are worth considerably less than those at Alacranes. In dry years there is enough land to make camp to fish the charal season, but the recent rise in lake level has inundated islote ranchos, precluding fishing.

Although the accord recognizes fishers' rights to continue fishing under the rancho system, they are reminded often of the incompleteness of that recognition and the uncertainty of the exclusive right. For example, whenever conflict arises among fishers or between fishers and the four restaurant owners who also hold island concessions, PESCA officers remind them of the limited scope and fragility of that recognition. They warn fishers that if they call attention to themselves through disputes, authorities (i.e., PESCA in Mexico City, SARH, or SEDUE) will say, "These people are only causing problems. To hell with them. Let's cancel their concessions and fine them for violating the law."

A second example of inconsistent recognition of fishers' rights to organize is the more general issue of allotting the area adjacent to a community to its organized fishers. Before the widespread adoption of outboard motors, most fishers set their gear within a couple of hours' rowing time from their community. This techno-logical constraint created de facto boundaries. The introduction of motors in the 1970s, however, removed this constraint on fishers' range. This, together with the increased number of fishers and gear, has brought into question what rights and preferences, if any, local fishers have over "outsiders" and adds to the problem of boundary definition.

Local PESCA staff members frequently receive complaints from fishers (both those with and without motors) about intru-sions by outsiders into areas they customarily fish. Whereas federal and state authorities maintain that the lake is free, that is, open access to any fisher with a permit, a phrase echoed by many fishers, the Chapala office has dealt with the complaints by suggesting that fishers ask permission to fish within 500 meters of another community's shoreline, thereby giving limited recognition to community preference over adjacent waters. If fishers deny access to outsiders, however, they are reminded that the request for permission was a courtesy and that the lake is, in fact, free to all who have permits to fish. Although this procedure has proved a viable mechanism for resolving conflict among fishers in some communities, it has not worked in San Pedro Tesistán. "Outsiders" come from Michoacán, where recent loss of fishing area related to the drop in the lake level has forced them to seek alternative fishing grounds. Despite PESCA officers' suggestion that they set their gear beyond the 500-meter "community zone," these fishers continue to fish nearby, often running over local fishers' gear as they motor to and from their gear. The limit of the local PESCA officers' enforcement power to within Jalisco is most apparent here, as they cannot force Michoacán fishers to comply with their suggestions. Still, San Pedro fishers expect the authorities' assistance, especially when their gear is damaged by outsiders' motors.

A third example of incomplete recognition concerns definition of who may and may not use the resource, again a matter of boundary definition. Although the local PESCA chief defers decisions to admit new members to the unions and cooperatives, he takes little or no action on their reports of noncomplying and nonpermitted fishers. To many fishers, the de jure state-property fishery, which is supposed to be accessible only to members of fishers' organizations, has become de facto open access (Bromley 1990). Fishers are frustrated by inconsistent recognition of their efforts to define user boundaries and assist monitoring of resource use. Their sentiments are reflected clearly in Paré's study, where fishers say to PESCA, "Either unite us or leave us free" (1989:107).

A final example of incomplete recognition is evident in PESCA's resistance to fishers' efforts to form coalitions. As mentioned earlier, San Pedro fishers had joined a coalition of fishers' organizations, known as Organizaciones Unidas del Sur del Lago de Chapala Marcos Castellanos (OULSCH).[16] OULSCH was sponsored by a private, nonprofit organization, Educación y Desarrollo del Occidente (or EDOC), that has tried to foster collective action and institutional development among fishers since PESCA first pushed for group formation in 1984. EDOC has encouraged fishers to analyze problems of resource use and work collectively toward their resolution (Paré 1989). In the late 1980s, OULSCH members pressured government agencies (PESCA, SARH, SEDUE) to take action to stop deterioration of the lake's water quality and quantity. The agencies retaliated against that pressure by fighting the coalition.[17] PESCA has refused to recognize the coalition and its member organizations. In 1990 the San Pedro union dropped out of OULSCH because of pressure from PESCA, including threats to cancel future aid and withhold permits. As problems with outsiders persist, however, union members perceive a lack of recognition by PESCA authorities of their problems and needs for assistance and are considering renewing their membership in OULSCH.

Summary and Conclusions

Local institutional development in the Lake Chapala fishery is notably beset by a lack of clearly defined boundaries and by ambiguities concerning fishers' rights to organize. The lack of clearly defined boundaries is found in the variation among legal prescription and government policy at three levels (federal, state, local), as well as group policy. Action by the Fisheries Secretariat is constrained at times by other agencies' (e.g., SARH, SEDUE) regulations and policies. The inconsistencies among these boundary definitions have emerged in the interpretive process necessary for the application of the law to local conditions.

Despite legal definition of the resource as state property, it has been portrayed to users as a common pool, to be accessed exclusively by members of fishers' organizations. Yet, authorities' failure to enforce the law and unify policy has created an open access situation. Fishers are not assured of an exclusive right to the resource and therefore have little incentive to maintain institutions to coordinate resource use, let alone devise new ones. One exception has been the rancho system. Although its status as a common pool system with its own set of institutions is uncertain, the lack of enforcement by authorities of land use regulations, to date, has allowed it to continue. Still, further institutional development to accommodate changes in the resource and its use is hindered by a lack of official support for rancho fishers' efforts to form their own organization.

There is a strong sense among fishers, both from the organizations studied and from other groups, that under present conditions, cooperation to resolve problems associated with their common resource use would be fruitless. There is no assurance that current rules will remain in place or be enforced consistently. The uncertainty of the resource use structure has forced fishers to elect what amounts to the dominant strategy in a prisoner's dilemma, whereby each seeks to maximize his own short-term gain because he has no assurance that others will cooperate in efforts to coordinate resource use. If boundaries were clearly defined and fishers' efforts to devise their own institutions for coordinating resource use were recognized by authorities, a number of social dilemmas confronting fishers might be resolved.

At Lake Chapala, fishers could benefit by establishing group fishing zones to reduce potential conflict and better monitor each other's activities. Rancho fishers could coordinate their commercialization activities to overcome price collusion by buyers. Fishers, in general, could improve upon poor marketing conditions throughout the region through the creation of stable coalitions. Yet, without assurance of their exclusive claim to the benefits of such activities through clear definition of resource and user boundaries and recognition by authorities of their rights to organize, such institutional development is unlikely.

From the Chapala case, it is apparent that definition of resource and user boundaries and authorities' complete recognition of resource users' rights to organize are conditions that must be met whether the resource is to be managed by fishers, by authorities, or by both in concert. The fulfillment of these two conditions is a function of the situational factors. At Chapala, these factors work together to obstruct boundary definition and to promote inconsistencies in authorities' acknowledgement of resource user traditions (e.g., the rancho system), needs (e.g., protection from gear destruction), and ideas (e.g., community fishing zones).

The rancho system, in particular, is a folk management system that could assist the rationalization and sustainable use of the charal fishery. The local PESCA officer's interest in it, and his efforts on its behalf, constitute one step toward its protection and enhancement. Without complete recognition by all interested authorities, however, the system's future remains in question.

The more recent institution, the fishers' organization, also offers opportunities and challenges in terms of local resource management. The inherent difficulty of bounding aquatic settings does not preclude or make problematic the specification of who may or may not use a given resource. Lake Chapala's unions and cooperatives offer a mechanism for defining and controlling resource use. Again, however, inconsistencies in specification and recognition of human boundaries relative to the resource, and the limited role such organizations are permitted in resource management, render them ineffective, if not destructive. Given fuller recognition, user group organizations could facilitate user definition and subsequent monitoring and enforcement activities, whether by management authorities or fishers themselves. In the Chapala case, these organizations are the result of both locally generated and externally mandated institutional development. As localized units, they generate a social milieu within which constraint of resource use by their members may be more effectively imposed. One risk of devolving some management authority to local organizations is that local elites or other powerful actors may take over the organization and preclude coordination of resource use among users and with external authorities. The union or cooperative, however, can counteract such tendencies as a democratic forum for decision making among resource users. Here authorities could work with fishers to insure that outcome.

Until the cultural and political constraints to clear boundary definition and recognition of these folk systems are removed, institutional development will continue to be stymied. More importantly, perhaps, the deterioration of these locally generated and hybrid mechanisms for resource management will be accompanied by the loss of a once-productive fishery resource.

Lessons for Modern Fisheries Management

1. Boundaries of resource use can be created, even if the resource itself is unbounded. Boundaries enhance fishers' sense of control over the shared resource and the likelihood that they will work to sustain its use over the long term.

2. Ambiguities and inconsistencies in law, policy, and their interpretation can add to the environmental and market uncertainties that confront fishers and discourage fishers in their efforts to manage the resource. If fishers are unsure if, when, and how a given rule will be supported or enforced, they have no incentive to abide by it.

3. Passive recognition seldom is sufficient to allow common-property resource institutions to function. Fishers must receive proactive support from governmental authorities.

4. User group organizations mandated by government cannot be assumed to possess the social or institutional capital sufficient for spontaneous institutional development. Authorities may have to act as facilitators and advisors to encourage such groups.

5. Where groups have devised their own institutions, they occasionally will require assistance from authorities to enforce their rules. Thus, boundary definition with active recognition of the fishers' role in common-property resource management is critical to the survival of folk management systems and to the

development of new local institutions for common pool fishery management.

Notes

1. The design principles include: (1) clearly defined boundaries, (2) well-fitting rules, (3) collective choice arrangements, (4) monitoring, (5) graduated sanctions, (6) conflict-resolution mechanisms, (7) minimal recognition of rights to organize, and (8) nested enterprises.

2. Three PESCA offices have jurisdiction over the lake's fisheries. The Chapala jurisdiction is the largest, with authority over 43 percent of the organized fishers and (supposedly) 60 percent of the lake. The Ocotlán, Jalisco, and Sahuayo, Michoacán, offices, respectively, have authority over 40 percent and 17 percent of organized fishers and 30 percent and 10 percent of the lake. The delineation of these jurisdictions is somewhat arbitrary, as neither fishers nor fish respect such boundaries.

3. Sources estimate that there are anywhere from 900 to 3,000 unregistered, or "free," fishers (Paré 1989).

4. Each species has its high and low production and consumption periods. Nonselective gear, such as traps and nets, enable fishers to catch throughout the year. The atarraya and mangueadora are species-specific and are used to best effect during high-production periods of January through April and July and August.

5. In this chapter, the terms *rancho* and *rancho system* will be used to refer to this fishery.

6. One buyer, who is also a fisher, takes the catch to Guadalajara by bus to sell it. The other, a former fisher, runs a stall in the nearby Jocotepec market.

7. This shoreline has retreated and advanced as human and natural forces have withdrawn and added water.

8. Some species are reserved for fishing cooperatives, but none of these species are found at the lake. Permits are required for three fisheries: (1) tilapia, carp, and catfish; (2) charal; and (3) whitefish (a prized but extremely scarce fish).

9. The Jalisco office is located in Guadalajara, with regional offices in Chapala and Ocotlán (at the northeast end of the lake). The Michoacán PESCA delegation is based in Zamora; its office responsible for Lake Chapala is in Sahuayo, southeast of Lake Chapala.

10. The emergence of this policy is tied to the prior designation and subsequent development of PESCA as the fishery-management authority. A local cacique

formed an extensive organization of fishers to assure his dominance as the principal middleman in Chapala. Until 1984, fishers were assured protection from potentially abusive authorities. They were also assured opportunities for equipment, loans, and other types of aid. At the same time, the cacique exercised considerable control over local and state fishery authorities.

11. An additional cost in some unions is participation in local political events. The rationale is that political support today provides at least some assurance of help tomorrow if the need should arise.

12. The alleged free fishers actually had obtained membership in another group, which is based in another lakeshore community. The local PESCA office has been in conflict with this other group in connection with past efforts to organize south-shore fishers into a coalition. Toward the end of the study, the local PESCA office was trying to re-establish more congenial relations with this group.

13. This core had succeeded partially in redefining the group and resource use boundaries through the annual renewal of the cooperative's rancho concession at *el islote,* a sandbar adjacent to Isla de los Alacranes. Whereas ranchos held by Chapala union fishers are individually concessioned, those of cooperative members are part of a single concession of federal land, which is partitioned among dues-paying members. This is consistent with the law governing cooperatives (SARH 1938a, 1938b), which states that all means of production are common property of the group. The president of the cooperative, who represents the group in dealings with SARH, has redistributed some of the subdivided parcels to complying members as an additional means of redefining group membership and resource use.

14. A few members of the Chapala union are not fishers, but fish buyers and vendors. The union president is a former fisher and restaurant owner; the secretary has not fished commercially, but is one of two charal buyers/processors in Chapala, and the son of the union's founder.

    Many fishers join when it is convenient, as during Lent, when there are benefits supposedly to be gained from a high demand for fish and its resultant higher price. Their entry not only crowds the fishery, but floods the market and eliminates any profit windfall created by the higher demand for fish. In reality, the additional benefits to be realized during Lent are few. One fisher commented that there is actually less demand for fish during the six-week period. People tend to buy fish only on Fridays, when religious custom prohibits the consumption of meat.

15. By the end of fieldwork in late April 1992, the list of twenty-seven Chapala union members to be dropped from the group had not been passed on to the PESCA office.

16. OULSCH espouses a three-point platform emphasizing ecology, production, and commercialization. Its activities have included latrine building, lakeshore clean-up, information campaigns, press conferences, and other tactics to raise lakeshore residents' consciousness of their interdependence with the lake and to pressure government into taking action to correct environmental abuses. This latter objective is a chief source of annoyance to PESCA, SEDUE, and SARH. It is responsible, in part, for the pressure felt by San Pedro fishers to drop out of the group.

17. The agencies (PESCA, SARH, and SEDUE) went so far as to accuse EDOC of being linked with the CIA and French intelligence in an alleged scheme to destabilize the Mexican government (Sandoval Lara 1990).

References

Bromley, D. W.
    1990   The Commons, Property, and Common Property Regimes. IASCP paper, September 27–30. Durham, NC: Duke University.

Cass, R. C., and J. J. Edney
    1978   The Commons Dilemma: A Simulation Testing the Effects of Resource Visibility and Territorial Division. *Human Ecology* 6:371–386.

Departamento Forestal de Caza y Pesca
    1939   *Código de Pesca.* México, DF: DAPP.

Edney, J. J.
    1981   Paradoxes on the Commons: Scarcity and the Problem of Equality. *Journal of Community Psychology* 9:3–34.

Hardin, G.
    1968   The Tragedy of the Commons. *Science* 162:1234–1248.

Jorgensen, D. O., and A. S. Papciak
    1981   Effects of Communication, Resource Feedback, and Identifiability of Behavior in a Simulated Commons. *Journal of Experimental Social Psychology* 17(4):373–385.

Lawry, S. W.
    1990   Tenure Policy Toward Common Property Natural Resources in Sub-Saharan Africa. *Natural Resources Journal* 30(Spring):403–422.

LeVieil, D. P.
    1987   Territorial Use Rights in Fishing (TURFs) and the Management of Small-Scale Fishing: The Case of Lake Titicaca (Peru). Ph.D. dissertation, University of British Columbia, Vancouver.

McCay, Bonnie J., and James M. Acheson, eds.
    1987   *The Question of the Commons: The Culture and Ecology of Communal Resources.* Tucson: University of Arizona Press.

McGoodwin, James R.
   1990   *Crisis in the World's Fisheries: People, Problems, and Policies.* Stanford, CA: Stanford University Press.

Oakerson, R. J.
   1986   A Model for the Analysis of Common Property Problems. In *Proceedings of the Conference on Common Property Resource Management,* National Research Council. Washington, DC: National Academy Press, pp.3–30.

Ortiz Martinez, J. M.
   1989   Contribución al Estudio de la Pesquería del Charal *Chirostoma* Spp. en la Laguna de Chapala, Jalisco, México. Tésis Profesional. Universidad Autónoma de Nayarit Escuela Superior de Ingeniería Pesquera, Nayarit.

Ostrom, E.
   1990   *Governing the Commons: The Evolution of Institutions for Collective Action.* Cambridge, England: Cambridge University Press.

Panayotou, T.
   1983   *Territorial Use Rights in Fisheries.* FAO Technical Paper No. 228. Rome: FAO.

Paré, L.
   1989   *Los Pescadores del Lago de Chapala y la Defensa de Su Lago.* Guadalajara, México: ITESO.

PESCA (Secretaría de Pesca)
   1990a Determinación del Potencial Acuicola de los Embalses Epicontinentales Mayores de 10,000 Hectáreas y Nivel de Aprovechamiento: Lago de Chapala, Informe Final. Guadalajara, México: Biotecs, S.XX, S.A. de C.V.
   1990b *Ley Federal de Pesca.* México, DF: PESCA.
   1987   *Reglamento de la Ley Federal de Pesca.* México, DF: PESCA.

Pinkerton, Eveyln W., ed.
   1989   *Co-Operative Management of Local Fisheries: New Directions for Improving Management and Community Development.* Vancouver: University of British Columbia Press.

Runge, C. F.
   1986   Common Property and Collective Action in Economic Development. *World Development* 14(5):623–636.

Sandoval Lara, Z.
   1986   [untitled] *El Occidental,* 6 agosto, p.1.

SARH (Secretaría de Agricultura y Recursos Hidráulicos)
   1989   *Ley Federal de Aguas y Su Reglamento.* Diario Oficial, 13 enero.
   1938a *Ley General de Sociedades Cooperativas.* Diario Oficial, 15 febrero.
   1938b *Reglamento de la Ley General de Sociedades Cooperativas.* Diario Oficial, 1 julio.

2

"Nowadays, Nobody Has Any Respect": The Demise of Folk Management in a Rural Mexican Fishery

James R. McGoodwin

Until fairly recently, South Sinaloa was the site of Pacific Mexico's most productive inshore fishery. The region's labyrinthine system of inshore lagoons and tidal estuaries provided ideal habitats for the production of shrimp, large fin fish, and oysters. All of these marine species were quite abundant, and the basis for valuable commercial harvests, until they became overexploited in the 1960s and 1970s. Since then, the fishery has remained tragically depleted.

My research in this region began in 1971 and has continued intermittently to the present (see McGoodwin 1976, 1987, and 1989). My long-term association with South Sinaloa, its fisheries, and its fishing peoples has been as much personal as it has been professional, with several friendships established there now spanning more than twenty years.

Most of my work in South Sinaloa has been an attempt to understand human articulations with the region's fisheries, beginning with the region's first human inhabitants through the present, but with particular concern for these fisheries' development during the modern era.

Folk Management During the Fishery's First Stage of Development (1930s Through 1960s)

In the modern era, the fishery's first developmental stage can be characterized as one of rapidly increasing production and commercialization, a time of great optimism and economic growth — indeed, the fishery's golden age. During this first stage the fishery was transformed from an underexploited, production-for-subsistence fishery in the 1930s to a fishery greatly involved in the production of seafoods for export by the 1960s.

In the early 1930s, as part of the federal government's postrevolutionary economic reform programs, fishing cooperatives were established in the region to supply regional packing plants with valuable exportable seafoods such as shrimp and oysters. Certain inshore territories were set aside for the exclusive use of these organizations, and from the beginning their members, called *cooperativistas,* prospered. Moreover, because the federal government regulated practically every aspect of the cooperativistas' harvesting activities, localized folk management of their activities was, for practical purposes, nonexistent.

At the same time the cooperatives were formed, a minority of the region's fishers chose not to join and instead remained freelance fishers, called *pescadores libres.* For these free fishers, at least in the early days, the exclusive territorial rights of the cooperatives posed little problem. Marine resources were abundant, and there were many productive fishing sites, other than those reserved for the cooperatives, where free fishers could enjoy abundant harvests.

Among the region's pescadores libres a system of folk management arose that had the following features. First, the region's various fishing communities asserted that their members had preferential rights to fish in certain good spots close to their home communities.[1] Because there was a general perception among all fishers that there was an abundance of good fishing spots scattered throughout the coastal zone, most fishers felt little need to venture to spots distant from their own home communities. Thus, on those rare occasions when fishers from different villages arrived at the same fishing spot at roughly the same time, the fishers from the community more distant from that spot would defer to the fishers who were closer to home.

Demographic, geographic, and technological factors helped to ensure that this system of etiquette worked smoothly during the early days of commercial development in South Sinaloa's fisheries. Demographically, the various fishing communities had small populations, so there were not many competitors for the region's abundant marine resources; geographically, the communities were mostly distant from one another, making it unlikely that competing fishers from different communities would very often run into one another; and technologically, local fishers utilizing dugout canoes relied on estuarine tidal movements to take them to and from their preferred fishing spots, which confined their mobility to a fairly small territory. It was not until the 1960s, when outboard-powered craft became widely used in this region, that fishers could easily invade the fishing spots that were informally claimed by fishers living in distant communities.

Moreover, among fishers living in the same community, other conventions of etiquette were observed. Regarding who had rights to fish at certain good fishing spots close to home, for example, the general rule was first-come, first-served and then take turns whenever activity at a particular spot continued for a prolonged time.[2] And during shrimp season, nocturnal fishers agreed to space themselves far enough apart so as not to diminish the attractive power of their individual kerosene lamps, which were used to lure the shrimp. And, again, regarding particularly favored shrimping spots, the rule was first-come, first-served.

These various conventions of etiquette were referred to as *respeto* (respect, or courtesy) among the region's fishers, and the system effectively minimized conflicts among competitors, both from different villages and from the same ones. Given the demographic, geographical, and technological factors mentioned above, and considering the abundance of marine resources and sites for harvesting them during the fishery's early days of development, this informal system of etiquette ensured harmony among competing fishers. Fishers from the same village faced unfavorable social sanctions within their communities should they violate the norms of etiquette, and fishers venturing further from home and into territories claimed by other villages found it prudent to remain deferential when encountering strangers — who were fishing closer to home.

It should be noted that this system of folk management was instituted mainly to avoid or minimize conflicts amongst fishers themselves, rather than to limit overall fishing effort and control fishing mortality. Otherwise, there were apparently no prohibitions, taboos, or discouragements placed on the use of any type of gear or regarding any particular fishing methods. These observations reaffirm those of McCay (1981:5–6), who stated that "most known cases of indigenous fisheries management hinge upon the management of access to fishing space rather than levels of fishing effort."

Folk Management in the Second Stage of Development (1960s and 1970s)

In its second stage of development, lasting from roughly the late 1960s through the late 1970s, the fishery continued to produce shrimp at a high level, but the heretofore abundant oyster and fin fish stocks were severely depleted early in this period. After many years of sustained high harvests, the oyster stocks were decimated by a great flood in the late 1960s and were unable to recover due to continued high harvesting pressure and an influx of chemical pesticides washing into the marine ecosystem from the region's developing agricultural sector. And the region's once-large stocks of fin fish were quickly depleted with the widespread introduction of nylon nets and dependable outboard motors in the early 1960s.

Although the reduction in stocks of the larger, more valuable fin fish contributed to an overall decrease in the availability of food for the regional populace, local fishers — including cooperativistas and pescadores libres alike — found that these losses were initially offset by the increasing profitability of the export trade in shrimp. Not only did prices for shrimp continue to soar, but now that the large fin fish — their main natural predators — had been severely depleted, shrimp stocks actually seemed to increase for a while.

By the early 1970s, however, shrimp harvesting began to exceed the fishery's theoretical maximum sustainable yield, and marine biologists began to issue warnings that the shrimp stocks were threatened. Heeding this alarm, the federal government began to deploy troops in the coastal zone, mainly to deter poaching for shrimp in the territories reserved for the cooperatives. But these and other efforts proved futile in slowing the now-rapacious harvesting of shrimp by fishers of many types.

By this time the fishery was overpopulated with fishers — cooperativized, free, and otherwise. Not only had the resident rural population undergone a boom expansion, but scores of newcomers had streamed into the region to exploit its valuable shrimp resources. Competition over shrimp became fierce, pitting cooperativistas and government managers against the rural pescadores libres, as well as various other members of the rural populace who eagerly harvested the region's shrimp. Indeed, as this second stage of the fishery's development matured, competition for shrimp resources became so intense that, in the words of one writer, the fishery became "an economic war zone," with shrimp becoming the basis for an "eternal conflict" (Lozano 1972).

Now localized folk management took on an ugly form. The earlier, informal systems of etiquette, which had been observed among competitors from different villages, had quickly broken down after outboard-powered boats and nylon gill nets became widely used in the 1960s. The outboard-powered boats enabled fishers to quickly travel to distant fishing spots claimed by other villages, deploy their nets, reap their harvests, and then run for home. And the larger nylon gill nets, stretching like impenetrable fences across certain estuarine narrows, proved so effective that they quickly wiped out whole stocks of the larger fin fish, while cutting significantly into the migratory shrimp stocks as well.

As a result of this sudden increase in competition, the rapid depletion of heretofore important stocks, the breakdown of the earlier system of etiquette, and the general impoverishment of the rural populace in this region, competing fishers began to commit violent acts against fishers from other communities. Fishing gear was stolen or cut loose, especially when encountered during nocturnal fishing activity, and eventually there were exchanges of gunfire resulting in injuries and deaths. After that, fishers ceased ranging so widely, turning their attention back to sites near their home communities, heightening competitive pressures there.

Strains in the local communities were also heightened as many newcomers from other parts of Mexico moved in and took up fishing. Typically, these newcomers were adolescents and younger men who were not accompanied by other family members and who rented marginal dwellings from local villagers. Thus, they were regarded with hostility and suspicion by many of the more established fishers and remained outside the web of the communities' longer-standing social organizations. Many of these newcomers also seemed unaware of the local norms surrounding respeto in the fisheries, and even when they learned about them, often did not feel as compelled to heed them as had the community's more established fishers. In time, these newcomers helped to hasten the breakdown of local norms of etiquette that had heretofore brought some order to local fishing activity.

As South Sinaloa's rural population continued to grow and economic conditions became more desperate, acts of violence increasingly became the prevailing mode of folk management in the region's fisheries, both within communities and between different communities. Typically these entailed spontaneous, individual attacks on property and persons, which were usually carried out in a clandestine manner. And although such actions were illegal under various Mexican laws, they nevertheless had a rather predictable and systematic character that was understood by most fishers and that, in this sense, maintained some degree of order in the fishery — however negative the fishers' motivation.

As the shrimp fisheries continued to decline, however, local cooperativistas redoubled their pleas for government help to dissociate pescadores libres from the shrimp resources, which further increased tensions and conflicts between heretofore friendly community members, neighbors, and kinsmen. Furthermore, as shrimp prices continued to rise in both regional and international export markets, new opportunists came to catch shrimp illegally for sale into clandestine, or black, markets. Thus, a new and criminal element began to be seen in the local communities, which further heightened strains and suspicions among the more established community members and further tore the fabric of local social and economic organization among the fishing populace.

As a result of all these factors, the formerly tradition-bound, internally cohesive, and semi-isolated rural-coastal communities slid downhill into social and psychological atomism, and eventually any semblance of orderly folk management in the fisheries effectively disappeared.[3]

The Collapse of the Region's Fisheries During the Third Stage of Development (Late 1970s to the Present)

As the 1970s drew to a close, South Sinaloa's shrimp fishery collapsed and has remained in a depressed condition to this day. Shrimp production is now only a fraction of what it was at its peak, and the shrimp industry is no longer a cornerstone of the regional economy, having been surpassed in importance by agriculture and animal husbandry. Oyster stocks remain depressed as well and are targeted mainly by individual, subsistence-oriented fishers hoping to supplement their families' food supplies. Now the only fish stocks of any interest are mostly subadult specimens of the formerly large fin fish, various smaller fish species that until recently were of little interest, and many other varieties of marine life — all of which are now caught mainly for reduction to fish meal. "We take everything we can catch now," a fisherman told me on my last trip to the area in 1992, "and if we get some food for our families, that is about the best we can hope for. Otherwise, nothing we catch has much commercial value, and the only way we get any money from fishing is by converting our catches into fish meal."

As South Sinaloa's fisheries collapsed during the late 1970s and early 1980s, countless fishers — both long-standing and fairly new — either turned to other occupations or moved away. Now, in the early 1990s, many of the region's fishers are transients who target a wide variety of marine life by utilizing small-mesh gill nets and/or longlines baited with hundreds of small hooks. Fishing from outboard-powered boats with fiberglass hulls, these fishers can rapidly move up and down the coastline, either in their fast-moving watercraft or by transporting the boats in trucks along the coastal highways. They are able to capture any remaining aggregations of inshore marine life of practically any kind — mostly for eventual reduction to fish meal. Highly mobile and opportunistic in every sense, these fishers are seldom integrated into the social fabric of any particular fishing community, and indeed some hail from inland communities that are far from the coast.

Localized folk management now seems nonexistent. Not only are there no apparent systems of etiquette, such as were seen during the early development of the fishery, but now even aggressive actions among competing fishers seem to have subsided. Localized management would now seem practically impossible to institute in this chaotic socioeconomic milieu — hardly worth the effort, given the pervasive pessimism about ever seeing the more valuable marine resources return to their former abundance.

Discussion

South Sinaloa's rural fishing peoples never had adequate time to develop viable, traditional systems of folk management such as those seen in societies having longer-standing articulations with local marine resources (see, for example, Johannes 1978 and Klee 1980, concerning Oceania; Berkes 1987, concerning subarctic northeastern Canada; and Ruddle and Akimichi 1984, concerning Japan and Okinawa). During pre-Columbian times South Sinaloa had a thriving indigenous population, but those peoples disappeared as a result of the onslaught of Spanish conquest and colonization during the sixteenth and seventeenth centuries. After that, the region remained essentially unpopulated until new settlers began to repopulate it in the mid–nineteenth century. Thus, the region had only a pioneering economy when significant commercial development of its fisheries began in the 1930s, and then, scarcely forty years after that, its fisheries were almost utterly depleted.

Yet, even if this region's fishing peoples had had time to develop viable systems of folk management prior to the modern era, I doubt that these could have persisted through the phenomenal developmental changes seen in the twentieth century. Indeed, most of the viable systems of folk management that are described in the literature on fishing peoples are found in situations where there has been a high degree of social and economic stability and where development and change have proceeded at comparatively slow, less chaotic rates. Localized management institutions simply cannot persist when localized fisheries undergo a rapid transformation from a subsistence-and-regional-market orientation to a cash-and-export-market orientation. Nor can they persist in the face of rapid socioeconomic change that sees rural-traditional, residentially stable, highly integrated local societies undergo rapid population growth, economic structural change, and an invasion of new residents who bring with them new values and new aims.

Impressionistically, with the collapse of South Sinaloa's fisheries, a new mentality about fishing seems to have arisen among the region's local peoples. Whereas the region's fisheries were once regarded as an abundant and reliable resource, and a career as fisher was esteemed as an important attribute of individual and family self-identity, now the region's fisheries — and fishing as an occupation — are spoken of as "ruined," "worthless," "no good any more," "not worth the effort," and so forth.

Hence, there is now a sense of futility regarding attempts to institute local folk management. "There aren't many fish anymore, and there isn't much we can do about it because this isn't our fishery anymore," an elderly fisherman told me on my last visit. "There are so many strange people living around here now, and nowadays nobody has any respect for anything."

Lessons for Modern Fisheries Management

1. Viable systems of localized folk management have a better chance of persisting in regions where socioeconomic change proceeds at an even and comparatively slow rate and where the human populace that is devoted to fishing has been residentially stable for a long time.
2. Attempts by localized community members to assert control over fisheries close to their communities will usually be unsuccessful if those fisheries are already experiencing intense competition from outsiders and severe depletion of important marine resources.
3. A state of ongoing depletion in a fishery will prompt low occupational self-esteem among fishers, as well as feelings of futility regarding attempts to exert local control over those fisheries.

Notes

1. These community-based, territorial, and fishing-spot claims were quite similar to those described by Acheson (1972, 1975, and 1988) in his publications about the harbor gangs in Maine lobster-fishing communities.

2. In Teacapán, the local community in which I have spent most of my time in the field, nearby fishing spots have colorful and descriptive names, their origins stemming back to the early days of the fishery's development, and perhaps longer: for example, La Boca (The Mouth, where a major estuary empties into the sea); Punta de la Barra (Point of the Bar, near the point of a sandbar); El Téjas (The Texas, origin of this name unknown); Los Gringos (The Gringos, i.e., The North Americans, where several North American sportfishers once made a large catch of snook in the early 1940s); and El Caimán (The Crocodile, a fishing spot near a beach where crocodiles formerly sunned themselves).

3. See McGoodwin (1976), describing aspects of social, economic, and psychological atomism in a fishing community in this region.

References

Acheson, James M.
   1972   The Territories of the Lobstermen. Natural History 81(4):60–69.
   1975   The Lobster Fiefs: Economic and Ecological Effects of Territoriality in the Maine Lobster Industry. Human Ecology 3:183–207.
   1988   The Lobster Gangs of Maine. Hanover, N.H.: University Press of New England.

Berkes, Fikret
   1987   Common-Property Resource Management and Cree Indian Fisheries in Subarctic Canada. Pp. 66–91 in Bonnie J. McCay and James M. Acheson, eds., The Question of the Commons: The Culture and Ecology of Communal Resources. Tucson: University of Arizona Press.

Johannes, R. E.
   1978   Traditional Marine Conservation Methods in Oceania, and Their Demise. Annual Review of Ecology and Systematics 9:349–364.

Klee, Gary A., ed.
   1980   World Systems of Traditional Resource Management. New York: John Wiley and Sons.

Lozano, Salvador
   1972   El Camarón: Eterno Conflicto. Técnica Pesquera 5(58):23–32.

McCay, Bonnie J.
    1981   Development Issues in Fisheries as Agrarian Systems. Culture and
            Agriculture 11(May). Bulletin of the Anthropological Study Group on
            Agrarian Systems, no. 11. Urbana-Champaign: University of Illinois.

McGoodwin, James R.
    1976   Society, Economy, and Shark-Fishing Crews in Rural Northwest Mexico.
            Ethnology 15(4):377–391.
    1987   Mexico's Conflictual Inshore Pacific Fisheries: Problem Analysis and
            Policy Recommendations. Human Organization 46(3):221–232.
    1989   Conflicts Over Shrimp Rights in a Mexican Fishery. Pp. 177–198 in John
            Cordell, ed., A Sea of Small Boats. Cultural Survival Report 26. Cam-
            bridge, Mass.: Cultural Survival.

Ruddle, Kenneth, and Tomoya Akimichi, eds.
    1984   Maritime Institutions in the Western Pacific. Senri Ethnological Studies
            17. Osaka, Japan: National Museum of Ethnology.

3

Folk Management in the Oyster Fishery of the U.S. Gulf of Mexico
Christopher L. Dyer and Richard L. Leard

Introduction

The management of fisheries resources within the coastal waters of the United States is currently in a state of crisis (McGoodwin 1990). Resources are diminishing, and some fisheries are collapsing due to fishing pressure, overregulation, pollution, habitat loss, overcapitalization, and other factors. At the same time, management has inadequately addressed these problems because of a lack of flexibility in administrative protocol and organization, because of conflicts among agencies and various special interest groups in fisheries, and because of different worldviews between managers and users (Smith 1990). Solutions have often neglected the importance of incorporating the resource knowledge, perceptions, experiences, and behavior of users, that is, the "anthropology" of fishery-based populations and communities. Recognition and incorporation of the anthropology of fishing communities, including their indigenous resource knowledge, could best be accomplished by managerial empowerment of users and by proactive

The authors gratefully acknowledge the assistance of Dr. Mark A. Moberg, Department of Sociology and Anthropology, University of South Alabama, for his assistance in the preparation of statistical analyses, and Mr. David M. Donaldson, Gulf States Marine Fisheries Commission, for his work with the presentation of graphics data. Most heartfelt thanks to Ms. Cynthia D. Bosworth for her skills and dedicated efforts to insure the quality of this chapter.

participation of maritime social scientists in fisheries management (McCay and Acheson 1987, Pinkerton 1989, McGoodwin 1990).

The importance of indigenous resource knowledge has been known for some time (Johannes 1978, Klee 1980, Ruddle and Johannes 1985, National Research Council 1986, McCay and Acheson 1987); however, indigenous viewpoints have seldom been incorporated into management regimes. On the contrary, as resources have become threatened and users more acrimonious, management policymakers have become more conservative and inflexible (Ward and Weeks, this volume). An outcome has been the elimination of local resource control and further decline in the sustainability of fisheries worldwide (McGoodwin 1990).

A wide range of terms has been used to describe the indigenous viewpoint vis-à-vis resources, including "indigenous resource management," "optimal foraging," "enlightened predation," and "efficient harvesting" (Charnov 1973, Agnello and Donnelley 1975, Johannes 1978, Krebs and Davies 1978, Chagnon and Hames 1979, Klee 1980, McCay 1981, Stocks 1987, Pinkerton 1989, McGoodwin 1990). Besides these and perhaps a few additional studies in natural resource management, there are few examples of successful incorporation of indigenous knowledge into management, or of direct demonstrations of behavioral correlates and environmental impacts of cultural practices that are believed to have conservative functions (Sahlins 1968, Feit 1973, McDonald 1977, Martin 1978, Lewis 1982, Stocks 1989, Hames 1987).

This chapter examines the oyster fishery of the Gulf of Mexico and presents an indigenous model of resource utilization in this fishery that is termed folk management. Folk management is here defined as indigenous traditions that facilitate the sustainable utilization of renewable natural resources. Folk management controls and influences human patterns of resource utilization or human behavior in ways that directly enhance and sustain resource productivity.

In this chapter, folk management in the oyster fishery of the Gulf of Mexico is examined within the states of Florida, Alabama,

Mississippi, and Louisiana, and it is conceptually fused to a community model elsewhere described as the natural resource community (NRC) (Dyer, Gill, and Picou 1992). We demonstrate that a fusion of folk management practices and prevalent anthropogenic characteristics of NRCs has facilitated sustainable production. The specific impact of folk management on oyster production and production per fisher, and the differentiation of these effects from economic and environmental factors, is considered using landings data from 1961–1988 in each of the four states (see Tables 3.1 and 3.2). It is noted that data on production per fisher are derived from commercial oyster landings by licensed commercial fishers. These data do not reflect recreational or subsistence fishing, that is, part-time fishing.

Table 3.1 Oyster production (landings) in Louisiana, Mississippi, Florida (west coast), and Alabama, 1961–1988

Year	Pounds Harvested by Dredges		Pounds Harvested by Tongs[1]	
	Louisiana	Mississippi	Florida (west coast)	Alabama
	1,000 lbs.		1,000 lbs.	
1961	9,708	2,427	3,218	509
1962	9,386	1,483	4,929	443
1963	10,168	4,061	4,278	992
1964	10,766	3,952	2,768	956
1965	7,930	2,332	2,760	465
1966	4,581	1,226	4,121	1,233
1967	7,528	1,963	4,528	1,994
1968	12,874	3,373	5,281	1,144
1969	8,744	841	4,886	453
1970	8,370	508	3,560	274
1971	10,346	1,126	3,518	247
1972	8,754	598	3,217	1,069
1973	8,837	498	2,404	590
1974	9,837	246	2,650	733
1975	13,525	1,023	2,128	638
1976	12,151	1,296	2,579	1,236
1977	9,623	767	4,071	1,549
1978[2]	9,501	283	5,880	760

(continues)

Table 3.1 *continued*

Year	Pounds Harvested by Dredges		Pounds Harvested by Tongs[1]	
	Louisiana	Mississippi	Florida (west coast)	Alabama
	1,000 lbs.		1,000 lbs.	
1979	7,423	164	6,121	460
1980	6,298	0	6,753	55
1981	8,846	241	7,167	1,330
1982	12,501	1,053	4,742	1,497
1983	13,141	3,181	4,216	336
1984	13,517	1,330	6,602	477
1985	14,246	1,156	4,360	1,442
1986	12,533	931	2,057	946
1987	11,760	58	3,671	88
1988	13,001	99	1,314	103

Source: Compiled from data contained in Fisheries Statistics of the United States (various issues) and unpublished National Marine Fisheries Service data.
[1]Includes a small amount of production by other gear.
[2]Data from 1978 through 1988 are considered preliminary by the National Marine Fisheries Service.

Table 3.2 Oyster production (landings) per fisher in Louisiana, Mississippi, Florida (west coast), and Alabama, 1961–1988

Year	Pounds Harvested by Dredges		Pounds Harvested by Tongs[1]	
	Louisiana	Mississippi	Florida (west coast)	Alabama
	1,000 lbs.		1,000 lbs.	
1961	13,908	4,694	5,655	677
1962	10,789	2,908	6,004	756
1963	10,992	6,238	5,341	1,465
1964	10,691	5,598	3,517	1,404
1965	7,790	5,170	3,817	615
1966	4,599	2,926	6,034	1,589
1967	7,589	3,091	6,177	2,508
1968	12,696	5,530	6,967	1,782
1969	8,779	2,594	7,061	709
1970	8,182	1,511	5,290	511
1971	10,035	2,858	5,523	538
1972	8,582	1,466	5,776	2,274
1973	8,638	1,838	4,659	1,497
1974	9,827	1,491	5,521	3,215
1975	13,675	4,142	2,742	2,562
1976	11,740	4,645	6,448	4,292
1977	9,537	2,769	6,796	4,376

Year	Pounds Harvested by Dredges		Pounds Harvested by Tongs[1]	
	Louisiana	Mississippi	Florida (west coast)	Alabama
	1,000 lbs.		1,000 lbs.	
1978[2]	9,370	1,874	8,364	1,723
1979	7,285	659	7,348	1,217
1980	6,110	0	7,301	131
1981	8,899	913	7,489	3,970
1982	12,172	3,540	5,310	4,229
1983	12,999	4,310	5,205	790
1984	13,265	1,795	7,519	1,340
1985	14,022	1,663	5,088	3,155
1986	11,042	1,268	3,070	1,739
1987	8,770	360	4,412	320
1988	9,286	678	1,492	368
1961–1988 average	10,045	2,733	5,569	1,776

Source: Compiled from data contained in Fisheries Statistics of the United States (various issues) and unpublished National Marine Fisheries Service data.
[1]Includes a small amount of production by other gear.
[2]Data from 1978 through 1988 are considered preliminary by the National Marine Fisheries Service.

Natural Resource Communities and Folk Management

Dyer, Gill, and Picou (1992) proposed that the social and cultural resources are best characterized as NRCs (natural resource communities). They defined an NRC as a "population of individuals living within a bounded area whose primary cultural existence is based on the utilization of renewable natural resources" (p. 106).

The relationship between a community of users and its natural resources can be described in a cyclic anticipatory utilization model (see Figure 3.1). This model diagrams interaction between the cultural and biological cycles. The cultural cycle is divided into four phases: (1) preparation, (2) harvesting, (3) utilization, and (4) anticipation. Preparation includes all activities involving the readying of gear, identification of target areas, crew selection, and, if appropriate, training. The harvesting phase is the actual event of collection and preparation. The utilization phase involves all

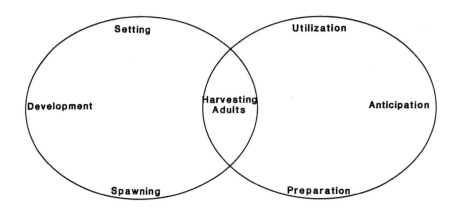

Figure 3.1 Anticipatory utilization model: the interrelationship of biological and cultural cycles of an oyster NRC.

activities associated with the conversion of the catch to usable resources. The final phase, anticipation, is an interim resting period during which predictions about the upcoming season and assessment of the previous harvest season guide decisions for future behavior in the fishery.

The biological cycle may vary depending on the fishery being studied; however, Figure 3.1 is a model for the eastern oyster, *Crassostrea virginica,* in the Gulf of Mexico. Setting of oyster spat is followed by development to the adult form, which spawns and is harvested at the temporal intersection of the cultural and biological cycles.

The cyclic connection between biological and cultural events is intrinsic to the production structure of an NRC, especially a fishery-oriented NRC. This structure is maintained by social networks based on kinship and cooperative ties. These networks allow for intense cooperation in the occupational roles of natural resource extraction, and they are maintained through transgenerational emphasis on fishing as a lifestyle. Dependence on natural

resources can limit occupational roles of NRC residents and intensify the assimilation of offspring (Firestone 1967).

Boundaries between outsiders and occupational residents are
fostered by the cooperative social networks of resource utilization
within the community (Acheson 1987). The strength of these
boundaries varies and leads to the general characterization of NRCs
as either open or closed. Boundaries are partly maintained by an
emphasis on "social contracts" in production, rather than social
relations. A social contract can be conceived as a voluntary and
mutual agreement to engage in purposefully limited, cooperative
endeavor. Furthermore, the degree of closure arises from variations in harvest technology and in the degree of economic pluralism that is prevalent. Open and closed NRCs conceptually
represent community models along a continuum of NRC types (see
Table 3.3).

**Table 3.3 Comparative sociocultural characteristics of closed and
open NRCs for the Gulf of Mexico oyster fishery**

Character	Closed NRC	Open NRC
Legal access to fishery	Unrestricted, or restricted by marine tenure rights or high fees	Unrestricted, no tenure rights, low fees
Access to utilization knowledge	Restricted along lines of kin associations and cooperative work groups	Unrestricted, but success often based on prior experience
Family orientation to fishery (traditional)	Traditional, transgenerational participation along kin lines	Opportunistic, varies with perception of catch success
Economic mobility of user population	Limited by occupational specialization in the fishery	Variable, depending on access to other opportunities
Reliance on folk management of resource	Traditional folk management with co-management	State management dominant, no co-management

In a closed NRC, a strong dependence on renewable, natural
resources limits occupational roles to resource extraction, processing, and related strategies. Occupational mobility is restricted by a

transgenerational emphasis on kin participation. Dependence on one or more related fisheries strongly links families to the locale. The establishment of relationships with resource marketers and processors, and the aquisition of knowledge regarding extraction sites and strategies require links with long-established social or kin networks.

In an open NRC, emphasis on social networks for a user to gain access to the fishery is not the same. Entrance or exit from the fishery does not carry as high a social cost as in a closed NRC. Open NRCs are more likely to present a variety of economic options in addition to traditional resource utilization, making opportunistic participation in any particular fishery common. Open NRCs thus have a higher frequency of "part-timers" utilizing common resources.

Folk Management and Its Relationship to Co-Management

Recognition of community-based roles in fisheries management has been identified in studies throughout the world, including Indonesia (Zerner 1991), the United States (Acheson 1988), the South Pacific (Johannes 1981, Iwakiri 1983, Ruddle and Johannes 1985), and Japan (Commitini 1966, Ruddle 1989). Common-property resource management is also well documented (National Research Council 1986, Runge 1986, McCay and Acheson 1987, Berkes 1989, Bromley and Cernea 1989, Herring 1990, Poffenberger 1990). The NRC model integrates common-property theory within a community-management system, and the reliance on renewable natural resources is the focal point of integration. This integration is conceptualized as folk management.

Because folk management is community focused, it results in stability of resource utilization and economic benefits to the community. As Ruddle notes, "the traditional knowledge handed down through generations of operators of resource systems assumes a potentially great role in modern management designs. . . . It has been realized that such local, unrecorded and not uncommonly encyclopedic traditional knowledge is often of great value in coastal zone management and in the design of resources systems" (Ruddle and Johannes 1985:2).

Co-management of fisheries implies "negotiated agreements and other legal or informal arrangements made between groups or communities of resource users and various levels of government responsible for fisheries management" (Pinkerton 1989:4). It has occurred in Canada and the United States, and one of the most successful cases of co-management occurs in the Lofoten fishery of Norway (Jentoft and Kristoffersen 1989). Like other international examples (Acheson 1975, Davis 1984, McGoodwin 1984, Berkes 1986), management of the Lofoten fishery is community based, spontaneously developed, and traditionally organized.

Co-management that incorporates folk management with state management has the potential for generating more sustainable production from fisheries than state management alone (McGoodwin 1990). State management policies, when enacted without knowledge or appreciation of folk management strategies that are already in place, can result in "state management extremism," a situation in which a fishery is conceptually defined as a fish stock or "resource" (Fox 1990).

Folk Management and Co-Management in the Oyster Industry of the Gulf of Mexico

The NRC structure characterized by the anticipatory utilization model (Figure 3.1) is presently found in all commercially linked oyster communities in the gulf; however, there are variances in the types of oyster NRC structures present in different areas and states. Open NRCs are less affected by folk management, and the associated communities lack formal or informal control of access to oyster resources. This open access allows harvesters to opportunistically enter or leave the fishery. Consequently, oysters in open NRCs are regarded as common property and are more subject to overharvesting. Mississippi and Alabama have oyster fisheries that we identify as open NRCs.

Oyster fishers' families are NRC linked, and an important part of identification as a legitimate user of area resources is kinship (Rockwood 1973). Achieving the status of insider takes several generations. Extended families are used to identify potential working partners in the fishery, and control of knowledge concerning the condition, location, and accessibility of resources follows kinship lines. In essence, kinship ties are used to establish legitimacy of individuals as community members (Rockwood 1973). Social outcomes of folk management in a closed NRC include self-reliance and an emphasis on community insularity. A strong demarcation prevails between interpersonal relations at the family and kin level versus those involving institutions outside of these relationships. Value is placed on communal integrity, and a lack of economic pluralism is replaced by a system of social reciprocity that is expressed through a sharing of labor, child-care, and income that results in communal security. As noted by Rockwood (1973), this system of interdependence gives coastal dwellers a sense of security that they cannot find anywhere else. Furthermore, his respondents were almost unanimous in stating that they wouldn't want to live anywhere else, and they often cited examples of persons who had done oystering in Louisiana or Texas for a season or two, but had then moved back. Louisiana and Florida have oyster fisheries characterized by closed NRCs.

An important characteristic of closed oyster NRCs is the effectiveness of folk management in limiting access to the oyster resources of a region. In the gulf, this access may be legally restricted (as in Louisiana) or controlled by social and cultural means (as in Florida). In either case, the oyster resources of the area obtain a property status. The establishment of this property status has a cultural foundation that reinforces and maintains a limited or controlled access system. This cultural foundation is folk management.

Specific predictions as to the effects of folk management in a closed oyster NRC are that the overall fishery production, production per fisher, and participation in the fishery will be stable over time. In an open NRC, overall fishery production, production per fisher, and participation in the fishery are predicted to be significantly more variable than in a closed NRC. These predictions are tested using ethnographic data and fishery statistics for comparisons of open NRCs in Alabama and Mississippi with closed NRCs in Florida and Louisiana.

State Folk Management and Closed NRCs in Louisiana

The oyster fishery and its management in Louisiana are the outcome of folk management practices operating within a cultural system of individual and family control of oyster reefs. Dalmatian oyster fishers arrived in Louisiana between 1840 and 1850 and can be credited with instituting folk management strategies of proprietary (territorial) control over oyster beds and cultivation of those beds. This heritage was common in the Adriatic region of Yugoslavia when these Dalmatian immigrants came to Louisiana. They succeeded in establishing sustainable and productive harvests and are credited with inventing modern oyster tongs and introducing oyster dredging to the Gulf of Mexico (Vujnovich 1974).

Traditional oyster camps served as temporarily closed NRCs. These camps maintained a proprietary, social control over oyster resources that through time evolved into the present leasing structure with its permanently closed NRCs. From 1850 to 1900, Dalmatian oyster families developed working relationships with state fishery managers and legislators and were instrumental in the enactment of the 1902 state leasing law. With this law came the establishment of a state management authority over the oyster fishery, the Louisiana Oyster Commission, which was actively supported by the oyster fishers. This dynamic interaction between

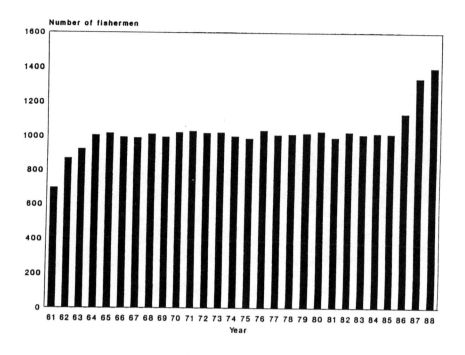

Figure 3.2 Number of dredge oyster fishers in Louisiana, 1961–1988.

folk management within NRCs and the state government to create an oyster leasing system in Louisiana is an example of co-management. In essence, folk management practices were incorporated into the legal regulatory structure.

Under the leasing system, oyster fishers pay for the privilege of working geographically defined bedding grounds, which produced approximately 80 percent of the total oyster harvest in Louisiana from 1962 to 1984 (Pawlyk and Roberts 1986). Practically all lease fishers in the gulf are found in Louisiana, where leasing has a long history. As with lobster fishers in Maine (Acheson 1988), oyster lease fishers invest time and resources in management as part of their production efforts. Those who work leased oyster beds are, in effect, following a tradition of oyster mariculture through reseeding of privately controlled areas. This tradition has helped

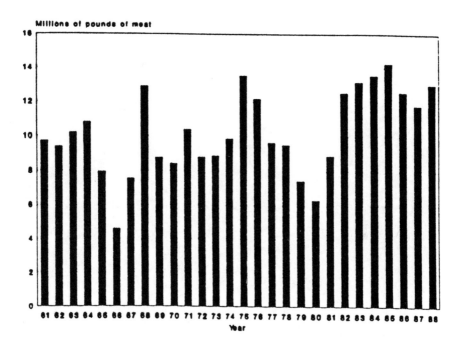

Figure 3.3 Oyster production by dredge in Louisiana, 1961–1988.

to make the Louisiana oyster fishery the most productive in the gulf.

Because leasing has such a long-standing tradition in Louisiana, the majority of the most favorable habitat is currently under private management. In contrast, a small number of nonlease fishers still actively compete for limited oysters in Louisiana, but they seldom have the extensive, cooperative networks common in oyster lease NRCs (key respondents in New Orleans; Berrigan et al. 1991).

The oyster-camp community network of families encouraged transgenerational participation in the fishery. Commitment to the fishery, measured by numbers of dredge fishers (primarily on leases), has remained high. Data from 1961–1988 show that participation was

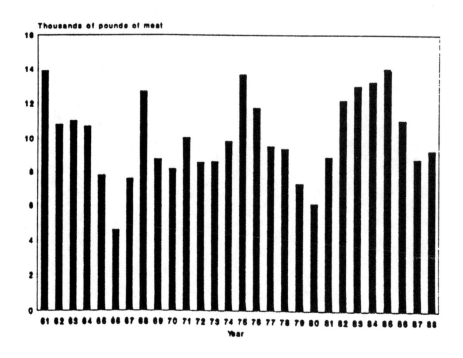

Figure 3.4 Oyster production per fisher by dredge in Louisiana, 1961–1988.

virtually unchanged during this period (see Figure 3.2). Although dredge production was somewhat variable in the late 1960s and late 1970s, it seldom varied more than 10 percent over the entire period (see Figure 3.3). Figure 3.4 shows that dredge production per fisher was also stable, and compared with other states, Louisiana maintained the greatest stability in production per fisher at the highest level.

State Folk Management and Open NRCs in Mississippi

The oyster fishery in Mississippi is made up of a variety of ethnic groups and cultures. Like Louisiana, Mississippi contains a large number of Dalmatian participants, but French, Spanish, and other European cultures are also present. In the late 1970s, Vietnamese and other Southeast Asian immigrants entered Mississippi and initially worked in the processing part of the industry. Today, Southeast Asian immigrants make up a large segment of the overall oyster industry in Mississippi.

The NRCs in Mississippi did not develop from the close, traditional interactions of families, as occurred in Louisiana's oyster camps. Mississippi has a small coastline compared to Louisiana, and traditional communication of natural resource information has occurred transculturally throughout the area. In addition, coastal Mississippians have historically maintained access to other resources such as timber, agricultural crops, livestock, and manufacturing. As a result there appears to be greater movement of individuals from oyster fishing families into other jobs, and a larger segment of part-time, opportunistic oyster fishers who oyster during the seasonal winter months and work at other jobs (for example, shrimping and construction work) in the summer. Mississippi has historically processed many times the amount of oysters available to be caught in the state, and processors have established and maintained business ties with fishers and dealers outside of the state in order to supply their market demands. These traditions have contributed to the openness of Mississippi's oyster NRCs.

The combination of these factors explains our open access characterization of oyster NRCs in Mississippi. As processing demands can be met by outside sources, controls on access would only serve to economically penalize local fishers and their families. Also, because other jobs are available in the area, the emphasis on transgenerational participation in families has been negatively reinforced, especially in years when supplies have been low. Finally, because the industry is processing oriented, the availability of information regarding oysters has been widespread, and because processors can handle more oysters than can be produced, competition for the resource has been high.

Oyster processors in Mississippi historically have practiced limited folk management by stockpiling processed shells and using their own vessels to replant public reefs without assistance from

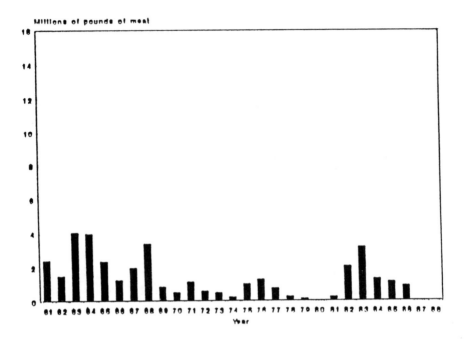

Figure 3.5 Oyster production by dredge in Mississippi, 1961–1988.

the regulatory agency. Historically, processors owned the vessels, and crews were hired to harvest oysters and plant shells; thus processors possibly exercised some control over harvests. The public reefs were, however, open to all fishers, and there was no limitation on access to reefs. Competition among fishers was merely for a job (captain, deckhand, and so forth), whereas the greatest competition was among processors, based on the size of their business, number of boats, and other factors.

Since the late 1960s, processors have divested themselves of oyster boats, and fishers now own most of the boats in the fishery. Competition among fishers and processors is still great, and processing capabilities still far exceed production.

Figure 3.5 shows annual oyster dredge production in Mississippi, and Figure 3.6 shows annual oyster dredge production per

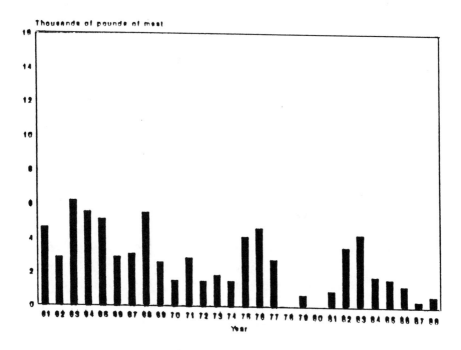

Figure 3.6 Oyster production per fisher by dredge in Mississippi, 1961–1988.

fisher for the period 1961–1988. Both production and production per fisher are quite variable and appear to be somewhat cyclic. This observation could be explained by cyclic overharvesting as a result of open access and a demand that exceeds the available supply.

State Folk Management and Closed NRCs in Florida

The oyster fishery in Florida that occurs primarily in the Apalachicola Bay area is characterized by closed NRCs. Fishers in these communities are descendants of Scotch-Irish immigrants who settled in this region. Rockwood (1973) described the nature of oyster families in the area as informal social organizations that are highly competitive with each other. Kinship ties define the boundaries of cooperation between family units. Oyster families represent production units, and the value of work and family cooperation in the fishery is an important factor that limits individual options to participate in other occupations. Although families encourage young members to increase their level of education and leave the community, their socialization has taught them values that draw them back. Also, there are limited opportunities for employment outside the oyster fishery.

Until recently, more than 90 percent of all residents in Apalachicola were directly or indirectly dependent on the local oyster fishery (Rockwood 1973, Stimpson 1990). Additionally, more than 92 percent of Florida's oyster production came from Franklin County in the Apalachicola Bay area (Sangree 1983).

Although competition is fervent among local families, access to the fishery for outsiders is highly discouraged. Cultural information concerning the location, extraction, and processing of oysters is not available to outsiders; however, community residents and marketers will take advantage of outside (state) resources when available (respondents; Weeks, personal communication). Because oyster fisher family members are limited in their ability to leave the oyster fishery and outsiders are discouraged from entering the fishery, access to oyster resources is communally controlled.

Fishery operations such as processing, marketing, and gear manufacturing and repair are locally controlled; however, not all processing and marketing networks are based on kinship. Patron/client relationships between oystering families and non-kin processors are common. These are maintained through loans to fisher-suppliers and a reciprocal loyalty of suppliers to their marketer-processors. Additionally, because fishers are very dependent on the sustainability of the oyster resources, they must actively folk manage them.

Folk management in Louisiana led to the development of co-management and a leasing system that has controlled access to oyster resources. In Florida, folk management has maintained the kinship-based units whose traditions and practices control access. Like Louisiana, folk management has also influenced government,

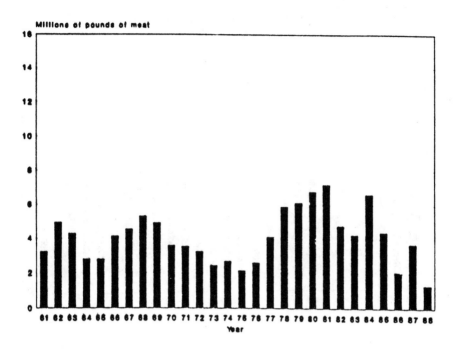

Figure 3.7 Oyster production by tongs in Florida's west coast, 1961–1988.

and the legal restriction that has arisen from this co-management is the limitation of gear to hand tongs. Tongs are more primitive and labor-intensive than dredges; a dredger can harvest more than ten times the oysters of a tonger on a given day. Consequently, this restriction serves to limit individual catches and access by dredge-oriented users.

Figure 3.7 shows oyster production by tong from Florida's west coast, primarily from Apalachicola Bay. Although environmental and economic factors have caused production to fluctuate over time, annual production has been relatively stable at approximately four million pounds of meat. Tong production per fisher in Florida has also been relatively stable at nearly 6,000 pounds per year (see Figure 3.8).

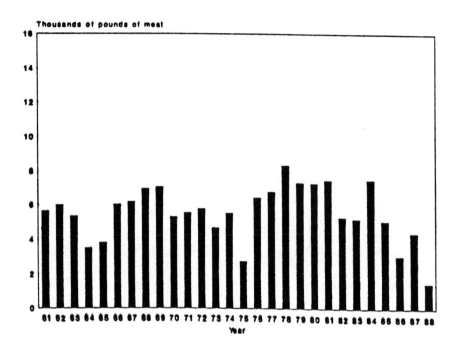

Figure 3.8 Oyster production per fisher by tongs in Florida's west coast, 1961–1988.

State Folk Management and Open NRCs in Alabama

Oyster NRCs in Alabama are characterized as open NRCs; however, Alabama oyster fishers are not devoid of kinship ties. Many fishers emphasize the traditional economic function of oyster fishing for their families.

Alabama's oyster reefs were worked by the state's indigenous inhabitants prior to the arrival of Europeans (Bryant 1972). Although independent tongers constitute the contemporary harvesting population, earlier commercial harvesting was more processor oriented. The first oyster factory, a floating plant in Bayou Coden, was constructed in the early twentieth century. The D and D (Dunbar and Dukate) factory in Portersville Bay also processed oysters and drew Polish and Bohemian workers from Boston and

Baltimore. This influx of multiethnic worker populations diluted ethnic cohesion, with the exception of French-Catholic settlements, such as those in Bayou La Batre.

Oysters have never been as commercially important in Alabama as in Louisiana. Shrimping, fin fishing, and crabbing have been additional components of seasonal harvest cycles for coastal oyster NRCs in Alabama. As in Mississippi, the comparatively recent influx of Southeast Asian immigrants has increased the ethnic diversity in contemporary communities, and many jobs in oyster processing are now held by Vietnamese and Cambodians.

A reliance on open access to oyster beds in Alabama has resulted in the evolution of traditions opposing privatization. Although oyster processors never actually controlled reefs, fear of this happening has caused independent oyster fishers to resist privatization of public reefs. Although leasing has been resisted by oyster fishers, some private oyster beds in the Fowl River drainage have demonstrated a sustainable productivity over the last several years. Here, harvests have exceeded the productivity of public reefs by an average of almost ten bags of oysters per working day (Van Hoose, personal communication).

Like Mississippi, Alabama has a small coastline and more loosely bound ethnic groups. Folk management has mostly consisted of individual cultch planting of reefs, and as in Florida the restriction of gear to tongs only. In addition, knowledge of oyster resources is widely shared. Alabama also has a broader economic base within the oyster community, and because cultural ties are less restrictive, individuals are better able to pursue other vocations.

Unlike Mississippi, Alabama's oyster industry is not as driven by the processing segment; however, the fishers themselves are quite mobile. For many years, when oyster resources have dwindled in Alabama, fishers have traveled to nearby Mississippi to tong oysters. Because Mississippi has traditionally relied on dredging as the primary means of harvest, Alabama tongers have met little competition. Another observation tends to support a protectionist behavior by Alabama tongers for their own reefs. Whereas Alabama

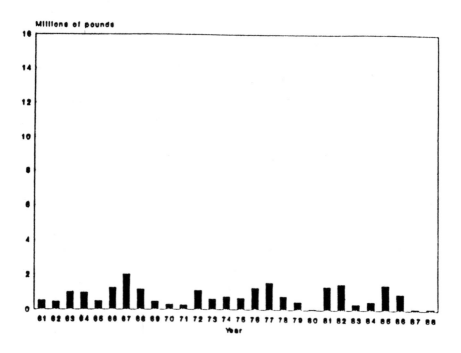

Figure 3.9 Oyster production by tongs in Alabama, 1961–1988.

tongers freely come to Mississippi when supplies are down or when Mississippi's supplies are up, Mississippi tongers complain that they are unable to do likewise. They fear having their boats burned and other reprisals; consequently, there are, in essence, no Mississippi oyster fishers operating in Alabama. This ability to move when resource abundance is low has created a situation where fishers are not dependent on local resources. Consequently, they do not actively practice extensive folk management to stabilize or sustain production from local reefs.

Figure 3.9 shows annual oyster production by tong in Alabama, and Figure 3.10 shows annual oyster production per fisher by tong for the period 1961–1988. Production averaged slightly less than one million pounds of meat per year, and production per fisher averaged approximately 2,000 pounds per year.

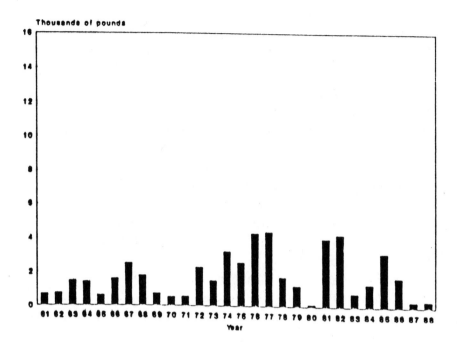

Figure 3.10 Oyster production per fisher by tongs in Alabama, 1961–1988.

Comparisons and Contrasts in the Oyster Industries of Florida, Alabama, Mississippi, and Louisiana

Mississippi and Alabama have relatively small coastlines, which are contiguous; however, oyster resources are influenced by completely different environmental factors. Fresh water bringing nutrients to reefs is the most critical factor in oyster growth and survival, and freshwater sources affecting oyster reefs in the two states are almost completely disjunct. Consequently, the abundance of oysters in these states is not affected by a common variable. Mississippi and Louisiana, however, have a common water source, the Mississippi River, that affects oyster abundance in both states.

As previously discussed, the oyster industries in Mississippi and Alabama are characterized as open NRCs, but folk management has been practiced and influences harvests. Fishers in both states are very mobile, and whenever oyster resources in Alabama are low, its fishers travel to nearby Mississippi to tong oysters and return to Alabama to sell their catch. Because Mississippi has traditionally relied on dredging as the primary means of harvest, Alabama tongers have met little competition. On the other hand, Alabama tongers may utilize a protectionist behavior with regard to their own reefs. Whereas Alabama tongers freely come to Mississippi when supplies are down or when Mississippi's supplies are up, Mississippi tongers complain that they are unable to do likewise; consequently, there are, in essence, no Mississippi oyster fishers operating in Alabama.

Similarly, Mississippi dredgers have historically harvested oysters in Louisiana and landed them in Mississippi. As most Louisiana fishers are lessees, they do not reciprocate and harvest from Mississippi.

This ability to move when resource abundance is low has created a situation where Alabama's and Mississippi's fishers are not dependent on local resources. Consequently, they do not actively practice extensive folk management to stabilize or sustain production from local reefs.

Discussion

To test for the effects of folk management on oyster harvests, we compared landings from Louisiana versus Mississippi and from Alabama versus Florida because of the commonness of the gear. Any bias in these comparisons would increase landings for Alabama and Mississippi, because these fishers frequently land oysters that have been caught from the adjacent state to which they are being compared.

When comparing production statistics to ascertain the presence of folk management effects, Figures 3.4, 3.6, 3.8, and 3.10 (production per fisher) are the most useful; however, actual production (Figures 3.3, 3.5, 3.7, and 3.9) and factors affecting production are also considered. In both Louisiana and Mississippi, the level of the Mississippi River is a key factor in the success or failure of the oyster crop. When the river is very low, high-salinity predators, primarily the southern oyster drill, *Thais haemastoma,* intrude on the major reefs in both states and devour the crop.

When the river floods, fresh water inundates reefs, and oysters may die. Flooding is particularly important in Mississippi because as floodwaters threaten levies and even the city of New Orleans, Louisiana, the Bonne Carre spillway is opened to divert floodwaters through Lake Pontchartrain and into the western portion of Mississippi Sound. Virtually all oyster reefs in Mississippi are located in this area, and flooding affects a significant portion of Louisiana's reefs in the eastern marsh area.

Major flooding and significant openings of the spillway occurred in 1973, 1979, and 1983. Figures 3.3 and 3.5 show that these events coincided with drops in production in both states, with the exception of Louisiana after the 1983 event. A common drop occurred in 1966, but actual causes are unknown. Environmental fluctuations resulting from the flow of the Mississippi River cause the trends in production and production per fisher for both states throughout the period (Figures 3.3, 3.4, 3.5, and 3.6). The key difference is in extent.

To test for the influence of folk management on oyster production and production per fisher in Louisiana and Mississippi, a three-part hypothesis was developed: (1) there is a significant difference in the variability of oyster yields, and states with closed NRCs exhibit more stability in yield over time than those with open NRCs; (2) there is a significant difference in the variability of oyster production per fisher, and states with closed NRCs exhibit more stability in yield over time than those with open NRCs; and (3) there is a significant difference in the mean values of annual oyster production per fisher, and states with closed NRCs have higher mean annual productivity over time than those with open NRCs. As most oysters are harvested by age three, a twenty-eight-year test for significance should eliminate the effects of environmental fluctuations.

Using an ANOVA test for variance, we found that Louisiana shows significantly less variation in dredging production than does Mississippi for the twenty-eight-year period (Figures 3.3, 3.5). The ANOVA test yields an R^2-value of 0.849 and an F-value of 302.77 that is significant at $p < .0001$ with 1 degree of freedom.

Using the ANOVA test again for variance in production per fisher, we found that Louisiana shows significantly less variance than does Mississippi for the twenty-eight-year period (Figures 3.4, 3.6). The ANOVA test yields an R^2-value of 0.7043 and an F-value of 128.65 that is significant at $p < .0001$ with 1 degree of freedom.

The difference in mean production per fisher is also significant, with mean production per fisher for Louisiana equal to 10,045 pounds of meat per year and Mississippi averaging 3,023 pounds of meat per fisher per year. A t-test for comparison of means gives a value of $t = |11.3|$ that is significant at $p < .0001$ with 54 degrees of freedom.

The average annual catch by Louisiana fishers was approximately five times that of Mississippi fishers. Biological/environmental and economic effects adequately explain the trends in production and production per fisher; however, the increased sustainability of catch, income, and profitability that are obviously realized by Louisiana fishers when their individual catches are so significantly larger can most acceptably be explained by cultural influences (folk management) that have limited access (Figures 3.4 and 3.6).

Reefs in Florida and Alabama are influenced by entirely different biological/environmental factors and perhaps by very different economies as well. No comparisons can be made in landings trends based on similarities of such effects (Figures 3.7 and 3.9). The same three-part hypothesis that was used to test for the influence of folk management and its effects on oyster production and production per fisher in Louisiana and Mississippi is used to compare these parameters in Florida and Alabama.

Using an ANOVA test for variance, Florida shows significantly less variance in oyster production by tong than does Alabama for the twenty-eight-year period (Figures 3.7 and 3.9). The ANOVA test yields an R^2-value of 0.6845 and an F-value of 117.18 that is significant at $p < .0001$ with 1 degree of freedom.

The ANOVA test for variance in production per fisher shows Florida with significantly less variance than Alabama for the twenty-eight-year period (Figures 3.8 and 3.10). The ANOVA test yields an R^2-value of 0.6356 and an F-value of 94.21 that is significant at $p < .0001$ with 1 degree of freedom. The difference in mean production per fisher is also significant, with Florida having a mean of 5,568 pounds of meat per tonger per year and Alabama with a mean of 1,776 pounds of meat per tonger per year. A t-test for comparison of means gives a value of $t = |9.7|$ that is significant at $p < .0001$ with 52 degrees of freedom.

Figure 3.8 shows that Florida's average annual production per fisher is nearly three times that of Alabama (Figure 3.10). Alabama's four highest production years per fisher (1976, 1977, 1981, and 1982) were only slightly higher than Florida's four lowest years (1964, 1965, 1975, and 1986), with the exception of 1988 (Figures 3.8 and 3.10). These facts support the contention that folk management practices have been operating within the closed NRCs of Florida to limit access, and the result is a more stable and economically productive fishery.

The characterization of Gulf Coast oystering communities as closed and open NRCs allows for these comparisons of fishery sustainability. Closed NRCs (Florida and Louisiana) show highly sustainable total oystering productivity and productivity per fisher when compared to open NRCs (Alabama and Mississippi). The comparative mean values for productivity and productivity per fisher are statistically significant. ANOVA comparisons of variance are also significant, and closed NRCs show statistically higher sustainability (less variance) over the twenty-eight-year data period. In the closed NRCs of Louisiana and Florida, productivity is maintained by a traditional reliance on folk management strategies.

Conclusions

Folk management in the closed NRCs of the Gulf of Mexico oyster fishery leads to sustainability of production and production per fisher. This is consistent with findings for other shellfish fisheries, where privatization of clearly bounded resources (Agnello and Donelley 1975) or the immobility of the resources (Christy 1982) allows for sustainability of productivity. Environmental and economic fluctuations have influenced oyster production among the states in this study, yet the magnitude of difference between Louisiana and Mississippi and between Florida and Alabama indicates that factors other than biology and economics are responsible for the stability and success of oyster fisheries in Louisiana and Florida. We propose that folk management practices in these states have served to limit access, thus minimizing possibilities of over-harvesting and increasing economic benefits to individual fishers and their communities. There is a clear distinction between the operative effects of folk management in the oyster fisheries of Louisiana versus Mississippi and Florida versus Alabama. In Louisiana, folk management has evolved into a leasing structure and thus a co-management system. In Florida it exists as a social control mechanism that limits access to cultural knowledge of the resource and markets. The key similarity is that oyster management in Louisiana and Florida has developed from folk management practices in closed NRC structures.

As noted by McGoodwin (1990), many of the world's fisheries are currently in crisis. In many instances, the crisis has developed as a result of rapidly changing cultural attitudes and economies that have affected or threatened traditional use patterns. These changes and perceptions of future changes have caused users to become increasingly contentious, both among themselves and in their relationships with managers. This contention is brought on partly out of fear that the resource will be depleted, but more importantly because traditional cycles of utilization are being altered.

Major advances in the efficiency of fishing gear, including vessels, and a tremendous increase in the demand for fish and shellfish over the past fifteen to twenty years have obviously contributed to the crisis situation in many fisheries. We propose that another major contributor has been the destruction of traditional use patterns and folk management practices within fishing NRCs. These two factors may even be considered as a cause-and-effect scenario, but they are certainly not the only factors that threaten the viability of various fisheries.

To the extent that increased fishing effort or changes in fishing practices have harmed, or perceptually harmed, fisheries, the most notable cause has been unlimited access. Whenever demand for any natural resources, especially marine fish and shellfish, drastically increases and harvesting is made easy, the potential for overfishing is great. The potential for destruction of the folk management regimes that have historically governed utilization is equally great.

In the oyster fishery of the Gulf of Mexico, folk management practices in Louisiana and Florida have not been as affected by demand and continue to limit access to the resource in a sustainable manner.

In most fisheries of the United States, managers have been ineffective or apathetic in controlling access, due to their inability to recognize a common-use ethic. Identification of folk management practices within fishery NRCs is imperative for developing effective management programs. Folk management and co-management may be the best strategies for establishing conservative management programs to obtain optimum sustained benefits from fisheries on a continuing basis.

As McGoodwin (1990) points out, a major challenge in many fisheries facing crisis conditions is integrating indigenous resource management into state management. To meet this challenge, local (communal) management should be recognized as an acceptable and potentially legal option. Empowering users to participate in the formulation of regulations and to share responsibility for enforcement within common-property fisheries is crucial if folk management and co-management are to succeed.

As in Louisiana, a fusion of folk management with state management to achieve a co-managed resource system has been successful. To realize this for other fisheries and communities, greater utilization of social-scientific data is necessary, and a spirit of cooperation and trust between managers and users should be promoted. As shown with the oyster fishery of the U.S. Gulf of Mexico, user-assisted management in the forms of folk management and co-management increased the sustainability of the fishery and the benefits to the user populations. Discovering latent folk knowledge and folk management practices can result in culturally appropriate co-managed systems and the further development of theories of folk management. Folk practices can be modeled, and these models can be applied to fisheries in different parts of the world to achieve solutions to the problem of sustainability. As indicated in this volume, folk management does not function perfectly in all places and at all times; however, consideration of folk knowledge and folk management can be used to ameliorate crisis conditions in many fisheries of the world.

Lessons for Modern Fisheries Management

1. Managers should identify existing systems of folk management through the study of fishing traditions of users.
2. Natural resource communities should be empowered to have greater managerial control over their own local resources.
3. Managers should provide protection for existing systems of folk management by taking the necessary steps to codify such systems.

4. Managers and users should be educated about the benefits of folk management and co-management through workshops and other nonconfrontational forms of communication.
5. Managers and maritime social scientists should develop models from existing folk-managed and co-managed systems to test and apply in fisheries where no such systems exist.

References

Acheson, J. M.
1975 The Lobster Fiefs: Economic and Ecological Effects of Territoriality in the Maine Lobster Industry. Human Ecology 3:183–207.
1987 The Lobster Fiefs Revisited: Economic and Ecological Effects of Territoriality in Maine Lobster Fishing. Pp. 37–65 in B. J. McCay and J. M. Acheson (editors), The Question of the Commons: The Culture and Ecology of Communal Resources. University of Arizona Press, Tucson, Arizona.
1988 The Lobster Gangs of Maine. University Press of New England, Hanover, New Hampshire.

Agnello, R. J., and L. P. Donnelley
1975 Property Rights and Efficiency in the Oyster Industry. Journal of Law and Economics 13:521–533.

Berkes, F.
1986 Local Level Management and the Commons Problem. Marine Policy 10:215–229.
1989 Ed. Common Property Resources: Ecology and Community-Based Sustainable Development. Columbia University Press, New York.

Berrigan, M., T. Candies, J. Cirino, R. Dugas, C. Dyer, J. Gray, T. Herrington, W. Keithly, R. Leard, J. Nelson, and M. Van Hoose
1991 The Oyster Fishery of the Gulf of Mexico, United States: A Regional Management Plan. Gulf States Marine Fisheries Commission 24, Ocean Springs, Mississippi.

Bromley, D. W., and M. M. Cernea
1989 The Management of Common Property Natural Resources: Some Conceptual and Operational Fallacies. World Bank Discussion Paper No. 57. Washington, D.C.

Bryant, G.
1972 Bayou La Batre: The Town That Is Commercial Fishing. The Fish Boat (October):31–42.

Chagnon, N., and R. Hames
    1979   Protein Deficiency and Tribal Warfare in Amazonia: New Data. Science
           203:910–913.

Charnov, E.
    1973   Optimal Foraging: Some Theoretical Explorations. Ph.D. dissertation,
           University of Washington, Seattle, Washington.

Christy, F. T., Jr.
    1982   Territorial Use Rights in Marine Fisheries: Definitions and Conditions.
           FAO Fisheries Technical Paper No. 227. Rome, Italy.

Commitini, S.
    1966   Marine Resources Exploitation and Management in the Economic De-
           velopment of Japan. Economic Development and Culture Change
           14(4):414–427.

Davis, A.
    1984   Property Rights and Access Management in the Small Boat Fishery: A
           Case Study From Southwest Nova Scotia. Pp. 133–164 in C. Lamson and
           A. J. Hanson (editors), Atlantic Fisheries and Coastal Communities:
           Fisheries Decision Making Case Studies. Dalhousie Ocean Studies Pro-
           gramme, Dalhousie University, Halifax, Nova Scotia.

Dyer, C. L., D. A. Gill, and J. S. Picou
    1992   Social Disruption and the *Valdez* Oil Spill: Alaskan Natives in a Natural
           Resource Community. Sociological Spectrum 12(2):105–126.

Feit, H. A.
    1973   The Ethno-Ecology of the Waswanipi Cree; or, How Hunters Can Man-
           age Their Resources. Pp. 115–125 in B. Cox (editor), Cultural Ecology.
           McClelland and Stewart, Toronto, Ontario.

Firestone, Melvin
    1967   Brothers and Rivals: Patrilocality in Savage Cove. Memorial University
           of Newfoundland, St. John's, Newfoundland.

Fox, W. W.
    1990   I'm on the Side of the Resource. National Fisherman 71(7):44–45.

Hames, R.
    1987   Game Conservation or Efficient Hunting? Pp. 92–107 in B. J. McCay and
           J. M. Acheson (editors), The Question of the Commons: The Culture
           and Ecology of Communal Resources. University of Arizona Press,
           Tucson, Arizona.

Herring, R. J.
    1990   Rethinking the Commons. Agriculture and Human Values 7(2):88–104.

Iwakiri, S.
    1983   Mataguali of the Sea: A Study of the Customary Right on Reef and Lagoon
           in Fiji, the South Pacific. Memoirs of Kagoshima University Research
           Center for the South Pacific 4(2):133–143.

Jentoft, S., and T. Kristoffersen
  1989 Fishermen's Co-Management: The Case of the Lofoten Fishery. Human Organization 48(4):355–365.

Johannes, R. E.
  1978 Traditional Marine Conservation Methods in Oceania, and Their Demise. Annual Review of Ecology and Systematics 9:349–364.
  1981 Words of the Lagoon: Fishing and Marine Lore in the Palau District of Micronesia. University of California Press, Berkeley, California.

Klee, G. A. (editor)
  1980 World Systems of Traditional Resource Management. John Wiley and Sons, New York, New York.

Krebs, J., and N. Davies (editors)
  1978 Behavioral Ecology: An Evolutionary Approach. Blackwell Scientific Publications, Oxford University, Oxford, England.

Lewis, H. T.
  1982 Fire Technology and Resource Management in Aboriginal America and Australia. Pp. 45–67 in E. Hunn and N. Williams (editors), Resource Managers: North American and Australian Hunter-Gatherers. Westview Press, Boulder, Colorado.

Martin, C.
  1978 Keepers of the Game. University of California Press, Berkeley, California.

McDonald, D.
  1977 Food Taboos: A Primitive Environmental Protection Agency (South America). Anthropos 72:734–748.

McCay, B. J.
  1981 Optimal Foragers or Political Actors? Ecological Analysis of a New Jersey Fishery. American Ethnologist 8:356–382.

McCay, B. J., and J. M. Acheson (editors)
  1987 The Question of the Commons: The Culture and Ecology of Communal Resources. University of Arizona Press, Tucson, Arizona.

McGoodwin, J. R.
  1984 Some Examples of Self-Regulatory Mechanisms in Unmanaged Fisheries. FAO Fisheries Report No. 289, Supplement 2:41–61. Rome, Italy.
  1990 Crisis in the World's Fisheries: People, Problems, and Policies. Stanford University Press, Stanford, California.

National Research Council
  1986 Proceedings of the Conference on Common Property Resource Management. National Academy Press, Washington, D.C.

Pawlyk, P. W., and K. Roberts
  1986 Interrelationships Between Public and Private Oyster Grounds in Louisiana: Economic Perspectives. Louisiana Seafood Production Econom-

ics, Louisiana Sea Grant College Program, Louisiana State University Center for Wetland Resources, Baton Rouge, Louisiana.

Pinkerton, E. W. (editor)
1989   Co-Operative Management of Local Fisheries: New Directions for Improving Management and Community Development. University of British Columbia Press, Vancouver, British Columbia.

Poffenberger, M.
1990   Keepers of the Forest. Kumarian Press, West Hartford, Connecticut.

Rockwood, C. E. (editor)
1973   A Management Program for the Oyster Resource in Apalachicola Bay, Florida. Florida State University, Tallahassee, Florida.

Ruddle, K.
1989   Solving the Common-Property Dilemma: Village Fisheries Rights in Japanese Coastal Waters. Pp. 168–184 in F. Berkes (editor), Common Property Resources: Ecology and Community-Based Sustainable Development. Columbia University Press, New York, New York.

Ruddle, K., and R. E. Johannes (editors)
1985   The Traditional Knowledge and Management of Coastal Systems in Asia and the Pacific, pp.157–180. UNESCO, Regional Office for Science and Technology for Southeast Asia, Jakarta, Indonesia.

Runge, C. F.
1986   Common Property and Collective Action in Economic Development. World Development 14(5):623–636.

Sahlins, M.
1968   Notes on the Original Affluent Society. Pp. 85–89 in R. Lee and I. DeVore (editors), Man the Hunter. Aldine, Chicago, Illinois.

Sangree, L. C.
1983   Oyster Cultivation by Reef Construction. Completion Report, P.L. 88–309, Project No. 2–370–D. Florida Department of Natural Resources, Tallahassee, Florida.

Smith, M. E.
1990   Chaos in Fisheries Management. Maritime Anthropological Studies 3(2):1–13.

Stimpson, D.
1990   Apalachicola Bay: Development Overfishing Wipes Out Oysters. National Fisherman (January):24–28.

Stocks, A.
1987   Resource Management in an Amazon *Varzea* Lake Ecosystem: The Cocamilla Case. Pp. 108–120 in B. J. McCay and J. M. Acheson (editors), The Question of the Commons: The Culture and Ecology of Communal Resources. University of Arizona Press, Tucson, Arizona.

Van Hoose, M.
> Letter to authors, 9 November 1990. Alabama Department of Natural Resources, Marine Resource Division, Dauphin Island, Alabama.

Vujnovich, M. M.
> 1974 Yugoslavs in Louisiana. Pelican Publishing, Gretna, Louisiana.

Zerner, C.
> 1991 Community Management of Marine Resources in the Maluku Islands. Report prepared for the Central Fisheries Research Institute, Ministry of Agriculture, Jakarta, Indonesia.

4

Resource Managers and Resource Users: Field Biologists and Stewardship

William Ward and Priscilla Weeks

In the area of fisheries management, the decline of fish populations despite a history of state-mandated natural resource management has challenged anthropologists to apply their knowledge of indigenous resource management to problems besetting state-controlled resources. Similarly, state managers have been challenged to integrate anthropologically informed models of resource use into policy goals. Anthropologists have explicated indigenous knowledge about natural resources and indigenous management systems with an eye toward enhancing local control over certain resources and/or the related, but somewhat different, goal of incorporating indigenous knowledge into state management regimes (see Albrecht 1990, Pinkerton 1989, Polunin 1985, McCay 1988, Foster and Poggie 1992). The bulk of this work focuses on the knowledge of "natural resource communities" (Dyer, Gill, and Picou 1992:106) and serves to legitimize this knowledge by favorably comparing it to scientific models of resource management (Johannes 1981).

Despite a burgeoning literature on folk knowledge and local management, the successful incorporation of local knowledge[1] into resource management is still relatively rare.[2] This chapter approaches the "problem" of resource management by redirecting the ethnographic gaze from fishers to managers, exploring some of the

Research for this chapter was supported by National Science Foundation Award DIR-9002894.

possible reasons for regulators' seeming inability or unwillingness to accept local knowledge.[3] Using the oyster fishery on the Texas Gulf Coast as a case study, this chapter considers regulatory scientists as a social group and looks at this group's training, socialization into management culture, and knowledge about natural resources and natural resource communities.

Responsibility for regulating the oyster resource resides with a large state agency, the Texas Parks and Wildlife Department (TPWD). Within the TPWD, the Coastal Fisheries Division (CFD) handles monitoring and policy formulation.[4] How the CFD scientists view their responsibilities to the resource and the oyster industry, as well as how they view human nature, may present some useful perspectives on TPWD's management strategies and help to explain the slow acceptance of local knowledge in a co-management plan.

We interviewed CFD field biologists in the two Texas bay systems having oyster industries to get an idea of how they saw their role, what methodology they used to monitor the health of the resource, what consensus existed for what is known about oysters, and what their opinions were on a number of issues concerning their data, the oyster industry, and CFD management. Some interesting consistencies existed among most, and at times all, of the field biologists interviewed that went beyond scientific or factual knowledge and included stereotypes of the modal personalities involved with the oyster industry (i.e., the fishers), as well as pseudo-scientific knowledge concerning what affects oysters and to what degree. These statements are of primary interest because they were widely accepted as obvious truths, thus becoming a kind of "public knowledge" (see Gusfield 1981:37) among field biologists that clearly has an impact on how they see their positions relative to those they manage.

Although field biologists are not policymakers, as a group they will pass on their perspectives, and biases, both factual and philosophical, to current and future policymakers, who can be greatly influenced due to their reliance upon the field biologist for data and opinions concerning the resource and the industry.

Managing the Oyster Fishery

The CFD scientists associated with the oyster resource are divided into technicians and biologists, who are located in coastal stations, and the management team, which is located in the inland state capital. The former provide information to the latter about oyster resources in the bays to which they are assigned.

When turning over regulatory authority to the TPWD in 1985, the Texas Legislature outlined the following mandate: prevent depletion while achieving optimum yield; manage the resource using the best scientific data; promote efficiency in the industry; and enhance enforcement (Vernon's Texas Code Annotated, 1985).[5] Enforcement is handled by TPWD game wardens. Resource management is the responsibility of CFD biologists, and theirs is a difficult task, given their dual mandate to promote yet restrain the oyster industry, mediate user conflicts, and wrestle with the problems inherent in the available scientific data.

The frontline efforts in data collection, resource and harvest monitoring, and public relations for TPWD are performed by the field biologists in the coastal stations. The CFD is responsible for monitoring all commercial and recreational species. Field biologists are responsible for all regulated species; thus they are not necessarily oyster specialists. There are two monitoring programs: harvest monitoring, an accounting of total fish caught, and resource monitoring, which gathers long-term trend data that will help the department detect population declines. All other research is conducted at a coastal station specifically devoted to research. The two monitoring programs mentioned above are the primary resource activities performed by the field biologists. In reference to their responsibility to oyster resources, most of the field biologists restated the legislative mandate that their duty was to prevent depletion while achieving optimum yield. Most implied that that was the ideal, but on a practical level they were mostly trying to prevent depletion or slow its rate.

The depletion of oysters is arrested through regulatory action. Currently, the CFD regulates via the establishment of open and closed oyster seasons, closure of areas to fishing, daily maximum limits on amount of catch, and limits on the size of oysters harvested. Some areas are permanently closed to fishing due to their proximity to the coast or because of fear of contamination. Prior to the TPWD's regulatory authorization, leases were granted to individuals to allow removal of contaminated oysters to a clean water site for purging and then marketing. This lease system is still in effect, although no new leases are being granted.

Educational Background and Work Experience of the Field Biologists

The education and work experience of the field biologists demonstrates a rather homogeneous background for the group. In regard to their level of work experience in the related field of coastal fisheries, they fall neatly into two groups. One group has more than twenty years of experience; the other has approximately ten years of experience. For the first group, the range is twenty to twenty-five years; for the second, it is nine to eleven years. Nearly all of this experience for both groups was gained by working for the CFD. On average per participant, only two years of work experience was acquired outside the CFD, and much of that was acquired working for the National Marine Fisheries Service. Nearly all of this outside work experience was obtained by the second group, which has the least amount of total experience.

The participants interviewed worked out of the Seabrook and Palacious field stations. The Seabrook office is responsible for the Galveston Bay system, which contains the bulk of industry activity, and Palacious is responsible for the Matagorda Bay system, the second most productive oystering area. Among the participants, the highest field position held within the CFD was Biologist II, which has supervisory responsibilities and is the highest field position within their respective offices. Two of the participants are now in administrative positions within the CFD.

The educational level among all participants is very similar. All but one have bachelor of science degrees in biology or fisheries. The one exception has a bachelor of arts degree in biology. Approximately one-third have master's degrees or several hours of course credit toward a master's degree. Most have earned at least one of their degrees from Texas A&M University, excepting one participant from Kansas State University, one from Antioch College, and another whose academic affiliations are unknown. Of those interviewed, three no longer work for the CFD, and one is retired.

The definition of a fisheries education appropriate to regulation surfaces now and again in journals devoted to fisheries management. Generally speaking, there is a great deal of homogeneity nationwide in fisheries science undergraduate programs (Chapman 1979). The curriculum is almost exclusively concerned with fishery science courses that prepare students to do research (Hester 1979), and, typically, there are virtually no social science or management courses to prepare the students for their role in the regulatory environment (Donaldson 1979, Miller and Gale 1986).[6] This can be problematic, as fisheries science students employed in governmental regulatory agencies can spend the majority of their time attending to public relations, supervision of subordinates, and policy formulation (Hunter 1984). Their education does not supply them with the people skills they will require for the duties they will perform in state regulatory agencies. They must learn these on the job (Schoning 1984).

The field biologists interviewed come from this sort of educational environment. Most have not been formally trained in the areas of social science, policy-making, and management (the one exception is the coordinator for the resource monitoring program, who holds a graduate degree in environmental management, which is under a business rubric). Their insights into the human aspects of the problems they deal with come primarily from their work environment and the long-established views of other students of fishery science programs who have come before them. All of this suggests that the pattern for the development of knowledge concerning natural resource management has been set for many years. This social knowledge, whether accurate or not, is passed on to the uninitiated soon after their initial employment. Without educational experience in the areas of human science, policy formulation, and management, the mold is set for perpetuating perspectives that may operate contrary to the stated purpose of a state-managed program. This will be especially true if the learned perspectives are incompatible with developing new and more effective means of managing the resource. If the resource manager is adverse to the resource user based upon a learned perspective, that manager may not readily accept a management technique that values the knowledge and input of the resource user. Thus, even when potentially useful perspectives are forced upon managers by users, managers may react by classifying the intrusion as a political issue that must somehow be worked around. If this indeed happens, the efficacy of a management program will depend unduly on the experience and intuitive abilities of higher management personnel or on the infusion of different perspectives through the inclusion of management personnel with formal training or experience in areas other than fishery science.

CFD scientists come to view the bay through a managerial lens, both in their coursework, described above, and in their early years in the CFD. The term *steward* was often used by interviewees to describe their relationship to the bay. They see the CFD, and themselves, as frontline personnel who are responsible for the health of the bay system. In order to keep abreast of changes in the bay, field biologists count organisms and take water samples. They learn to relate to particular bay systems through observations made during these monitoring activities. Their observations are channeled into a preset format, the data sheet, which dictates what information is considered relevant and how that information should be collected.

Their role as stewards necessarily entails managing people, as human activities can have severe impacts on natural systems. On the other hand, none of the field biologists interviewed expressed what are commonly referred to as "preservationist" attitudes. They all favored consumptive uses of the bay, and some suggested they had a responsibility to help industry. But otherwise, all were quick to point out that human use must be regulated.

Theoretically, field biologists assigned to various bay systems have more contact with fishers than those on the management team, stationed several hundred miles away. According to our interviews, however, contact with regulated groups varied greatly. Several biologists commented that contact with the industry had declined over recent years, and some of the newer staff members who had been hired within the last ten years claimed to have little contact with the industry. Nevertheless, these biologists still had definite ideas about how fishers operated and what motivated them. These generational differences in degree of contact with fishers was noticed after our interviews were complete, and we can only guess how the newer biologists learn about fishers.

Fisheries courses often include references to Hardin's (1968) work on the tragedy of the commons. We speculate that this knowledge, which is learned as a scientific model of how fishers can be expected to behave, is reinforced by stories about the activities of fishers, both as individuals and collectively, and substantiated through the monitoring data, which (sometimes) portrays declining fish stocks. We were also told such stories during our interviews. We speculate that the monitoring program, designed in part by a statistician, has contributed to this separation from fishers. When the monitoring program was first introduced, biologists relied on fishers for information about reefs and used technology employed by fishers to gather the samples. By now, however, the reefs have been mapped and the monitoring technique statistically rationalized so that sites to be monitored are chosen randomly and no longer reflect where fishers fish.[8]

As it is now conducted, the program is a source of dispute, which further separates fishers from biologists. Bays and portions of bays were closed in the past when the CFD's sampling procedures indicated that the oyster population at certain harvesting sites had been depleted beyond a level that was "safe." The oyster industry is therefore concerned about the methodology that is now used to collect samples. Fishers criticize the equipment used, the areas sampled, and the ability of CFD personnel to catch oysters under the new monitoring program. They claim that the biologists are underestimating oyster abundance because they sample the wrong places (i.e., they should sample known reefs, not randomly selected grids representing bay bottom) and also because the biologists are not good harvesters.[9] The field biologists we interviewed acknowledged, at least in passing, that these criticisms indeed existed. Not all of them, however, discussed them in much detail, and all agreed that the sampling is needed and useful for long-term trend analysis. They also agreed that oyster fishers were better at catching oysters than they were but otherwise felt that this was not a salient issue because it was not the biologists' purpose to catch oysters for market.

Public Knowledge

Public knowledge, as the term is used here, is defined using elements treated in Joseph Gusfield's *The Culture of Public Problems* (1981:36–37). Public knowledge as compared with private knowledge is made up of "facts which are shared by many people in the society who have no, or very little, personal knowledge of individual cases making up the aggregated facts" (1981:37). Although he discusses public knowledge at the societal level, the concept works well here. A group with a common background, common training, and confronted with a common problem — resource depletion —

has developed a body of knowledge about the nature of that problem and the difficulties to be faced in overcoming it. This knowledge is not particularly personal, nor is it derived from firsthand experience. To be sure, some of it may have been confirmed by personal experience, but the strength of the opinions of the biologist participants in this study concerning the nature of the oyster resource and oyster fishers could not have been derived solely from personal experience, especially as, by their own statements, their level of interaction with fishers has been very low. This suggests that they subscribe to a set of "facts" that are mainly learned secondhand, rather than experienced.

Oysters

The above discussion applies to some of the "facts" concerning oysters and what affects them or their habitat. These facts may or may not be accurate. They do not, however, appear to be grounded in specific research projects; rather, they seem to be extrapolations from basic scientific knowledge. For example, the issues surrounding freshwater inflows and the destruction of habitat were discussed more as opinions or probabilities than as facts grounded in solid scientific research. Yet these were repeatedly asserted by nearly all we interviewed as significant factors affecting the health and vitality of the oyster resource.

Following is an analysis of the opinions held by the field biologists and managers working for the CFD whom we interviewed. Special attention is given to their commonly held opinions concerning their data, the industry they regulate, and the management philosophy they work under.

There were facts about oyster biology that were agreed upon by everyone we interviewed concerning what affects oyster mortality, how oysters spawn, their reproductive cycle, and what habitat they thrive in. These facts were derived from scientific research on oysters. Of less certainty, however, and less supported by scientific research, is the question of the primary factor affecting oyster abundance in bay systems.

Salinity, harvest pressure, predators, and pollution were commonly cited problems. Salinity, or more specifically, the lack of freshwater inflow, was considered to be the most important variable affecting oyster populations. All of the interviewees referred either to the detrimental effects of dredging channels[10] or to the lack of fresh water coming into the bays due to dams and reservoirs along the rivers, streams, and bayous that feed the bays. Control of these factors is under federal, state, or municipal authorities.

The destruction of habitat ranked second with our interviewees as an important variable affecting oyster populations. Interviewees offered different reasons why this destruction occurs. Some spoke of environmental pollutants' causing a chain reaction affecting the livable habitats, but more often they spoke of actual physical destruction due to dredging, construction, and harvesting. And a few of the interviewees linked overharvesting with reef damage, feeling that harvesters dredged reefs below the surface, causing them to silt over.

Building new habitats and restoring old habitats by replacement of shell onto reefs or bay bottom is a project the CFD is now investigating with the help of a federal grant. Nearly all of the participants talked to some degree about the need for shell replacement, about creating a substrate (called *cultch*) for spat (juvenile oysters) to attach to so the fishery could be enhanced and populations maintained or increased. The only participant who did not mention shell replacement was not directly asked about it. Those who elaborated on what methods might be used often commented on their anticipated lack of cooperation from the oyster fishers.

Fishers

All of the managers interviewed felt that it is best for the CFD to have a working relationship with the oyster industry and, specifically, oyster fishers. However, none of the them felt that the

CFD's regulatory power or final say should be turned over to the industry. Each participant in his or her own way and in varying degrees indicated that the CFD should maintain regulatory controls, such as size limits, bag limits, length of season, culling percentage, and ability to close a season or an area.

Their unanimity derives from two views held by all of the participants: first, that the purpose of the CFD is to maintain and enhance the fisheries through regulatory control and, second, that the oyster fishers "will take what they can until they can't take any more," "will fish until the last fish is caught," and "will take every last living one." Such statements were made by virtually every one of the participants. A few expressed the opinion that such actions were a part of human nature. It is striking that all but one of the participants said almost the same thing with the same emphasis, regardless of age or years of experience. The fact that this last view is firmly believed by members of the regulatory agency shows the weight of their perceptions.

Another statement repeated by all interviewees concerned what they believe oyster fishers know. Without hesitation all interviewees stated initially that primarily, oyster fishers know how to find, catch, and sell oysters. A few qualified that assertion by saying that some fishers knew more than others concerning the biology of oysters and what affects them. Usually these fishers were described as being either more educated than other fishers or in a position to profit from additional knowledge (i.e., leaseholders).

The Construction of a Public Problem: Depletion

The most contentious biological issue between fishers and field biologists concerned the ability of the oyster population to recover from a depleted state. Depletion was defined statistically by the CFD and identified through the monitoring program. However, in the mid-1980s three things happened in rapid succession that changed the views of biologists at the CFD concerning first the levels at which depletion is imminent and, second, the use of season closures as a regulatory tool in circumventing depletion.

The first occurrence involved a lawsuit the CFD lost to the oyster industry. The suit was prompted by a TPWD decision to close the 1987 season based on the CFD's resource monitoring data, which indicated a resource near depletion. The oyster industry felt that the TPWD was acting capriciously and filed a successful lawsuit that forced the department to open the season that same year. In the view of most field biologists interviewed, this changed management's desire to invoke closures as a regulatory tool, even though doing so is still within the CFD's power. Most participants implied that CFD's management team did not wish to chance another legal and political loss to the oyster industry.[11]

The second occurrence affecting the use of closures was the inclusion of the Oyster Advisory Committee (OAC) in CFD decisions. The OAC serves as a liaison between industry and the division. Generally, the existence of an advisory committee does not constitute co-management because regulations are usually constructed by the agency, and committee members merely have the option of approving or rejecting them (McCay 1988). Although it is true that the oyster management plan was written before the institution of the advisory committee, industry members do introduce agenda items through the committee, which they expect the division to act upon. In this way, the advisory committee has the potential to act as a vehicle for the introduction of local knowledge into day-to-day management, a theme that we will return to.

The third occurrence was the recovery of the San Antonio Bay oyster population after a major freshwater kill. It had been thought that San Antonio Bay was either permanently dead or would not be productive for many years. Yet monitoring data revealed that San Antonio Bay's oyster population returned to productive levels in two to three years. This demonstrated to many of the participants the surprising regenerative capability of oyster populations. The view now expressed by all of the participants is that oyster populations cannot be depleted to a point from which they cannot return, unless the cultch is removed. It is significant that fishers had long maintained that oysters cannot be depleted, that a reef can recover as long as a few oysters are left. The department was aware of this argument but did not grant it much credence it until its own monitoring data indicated that a bay had been able to recover. Local knowledge about the periodicity of abundance was therefore legitimized through scientific knowledge.

Smith discusses the way in which fisheries management is an articulation of a particular view of nature, namely one expressing "periodic order" and lineality (1990:4).[12] Lineality is an especially important concept in the depletion models that are mobilized in management to portray steady population declines. We are not suggesting that CFD biologists force nonlineal "natural" patterns into lineal "artificial" ones; rather, we suggest that differential social constructs of time are involved. Fishers commonly criticize depletion graphs for not portraying the cyclical nature of changes in oyster abundance that they have noticed during their tenure on the water. They claim that the snapshots of populations captured by yearly assessments cannot capture the long-term cyclical nature of oyster abundance and should not be used for determining bay closures because monitoring programs and the depletion graphs depicting the results portray only part of a longer cycle. Although the TPWD no longer uses monitoring data to close oyster reefs, citing the San Antonio Bay case as a justification, the department does believe in the possibility of absolute decline over time, even if it may look more like a downward spiral rather than a line.

Neither view of nature — the linear and ordered view embedded in scientific models nor the cyclical and less ordered one offered by fishers — is more correct. Both linear and cyclical patterns are cultural constructions. Therefore, by interpreting these charts as truly representing decline over time, biologists are reifying a cultural construction of time as captured through the monitoring program into a natural fact. To quote Barley, "one of the most potent techniques we humans have for turning culturally arbitrary behavior into social fact consists of our tendency to treat

even self-imposed temporal boundaries as inviolable external constraints" (1988:125). In short, the graphs utilized by the TPWD
become "reality" and pre-empt the more grounded "reality" of the
fishers.

Until the recovery of San Antonio Bay, the ability to deplete
the oyster resource was an accepted "fact" among regulators. It was
common sense that a sedentary and limited resource such as
oysters could be depleted.[13] The monitoring program provided
scientific evidence in the form of graphs depicting a fairly steady
(linear) decline in numbers of oysters, and in that way overharvesting became identified as a public problem. As Gusfield (1981) points
out in his discussion of the construction of the problem of the
drinking driver, the process of identifying and labeling a public
problem involves the exclusion (or ignoring) of other, alternate
constructions. The construct *drunken driver* excludes as problematic the quality of roads or cars and is embedded in assumptions
about individual (versus, for example, state or industry) responsibility. Similarly, the construct *overfishing* excludes as problematic
pollution, dredging, salinity, and disease, all of which were otherwise recognized by interviewees as contributing to declining oyster
populations, but over which the TPWD has no control.

The TPWD can control overharvesting. Thus, overharvesting
becomes the "passage point" (Callon 1986:205), the issue through
which the various actors interact. The ensuing struggle to define
the conditions of depletion and its relationship to overharvesting
structures the relationship between fishers and managers. If fishing
pressure played no role in depletion, there would be no need to
regulate the industry, except, perhaps, regarding matters of health.
It must be noted that the conditions that give rise to overharvesting
as passage point are embedded both in the different views of nature
held by the CFD and fishers, and in the very real constraints the
TPWD faces vis-à-vis other user groups as articulated in its mandate
to manage resources.

Gusfield (1981) discusses the notion of the "ownership" of
public problems. The owners of the problem of the drunken driver

are the members of the group Mothers Against Drunk Driving. This group has kept the issue in public view, has pushed for regulations, and has contributed to the definition of the problem by focusing attention on the driver, rather than on the contributions of bad roads, faulty cars, and so forth. Owners sometimes have responsibility for both the definition of the problem and its resolution. This is the case for the TPWD and the problem of depletion, stemming from the TPWD's mandate from the legislature.

The proprietary attitude of the CFD is manifest in myriad ways. CFD biologists defined the nature of the resource when they wrote the oyster management plan. The source document for the plan included both basic scientific information about oysters and data about oyster abundance that had been gathered through the CFD's resource monitoring program. When making decisions about day-to-day management, however, the CFD chooses to give priority to its monitoring data, which it views as far better empirical evidence on trends of oyster abundance than the more theoretical models offered by scientists in universities, some of whom are critical of CFD management.[14] Debates between the CFD and university scientists center on the lack of data about certain parameters deemed necessary for good modeling (e.g., reproduction and recruitment, and the relationship between hydrological conditions and spat set). Although the bulk of citations in the oyster management plan's source book are from academic scientists, some of whom question the ability to deplete oysters, it was the CFD's monitoring data that set the stage for management through limits, seasons, and so on (Quast et al. 1988).[15]

The proprietary nature of the CFD biologists' definition of the problem of oyster depletion is also manifest in their redefinition of the cause of depletion. After the TPWD accepted the notion that overfishing alone did not necessarily lead to depletion, it identified habitat degradation as the key problem and took action to restore oyster beds by obtaining grant money to place cultch on shrinking reefs. Thus, the TPWD, rather than university scientists or industry, retained the responsibility for identifying and then correcting 'situations that could lead to the depletion of oyster resources. In this way, the department still owns the problem of depletion.

Conclusion

Both groups — managers and fishers — identify oyster depletion as a public problem. Within their definition of the problem and its causes, both point to forces outside their control as having significant effects on the oyster population and the general health of the bay (e.g., pollution, dredging, and freshwater inflow). However, they then come to odds regarding the role harvesting has on what may be a long-term depletion of the resource.

The CFD's primary role with regard to oysters or any other bay species is that of a regulatory agency. CFD employees' training is grounded in fishery science and resource management, not social science and industry management. They are the legally authorized stewards of a public problem, which they are responsible for defining and correcting. And, as the CFD watches the resource decline, the only controls it can assert are to impose restrictions on the fishing industry. Hence, even when the historically relied-upon basis for regulation (the argument that overfishing leads to depletion) is called into question, another concept related to overfishing arises to take its place (the argument that overfishing destroys substrate), thus allowing the CFD to continue to use the same regulatory and information-gathering tools they have traditionally relied upon. In other words, overfishing is still the core conceptual framework for management, because the CFD is powerless to control other variables such as pollution, freshwater inflow, or dredging.

The CFD's focus on overfishing is legitimized through stories about fishers' values and inability to cooperate with one another, coupled with the notion that fishers know only how to find, catch, and sell oysters. The industry has been brought into the management arena through the Oyster Advisory Committee (OAC), but our analysis of transcripts reveals that the OAC is not used as a vehicle for management based on local knowledge, but rather for conflict resolution.[16] And even if the OAC were an arena for the incorporation of local knowledge into management, it still would not be able to address some issues that are identified by both parties as central to depletion.

Pinkerton (in press) discusses long-term planning as an integral component of resource management. For the past three years, the Environmental Protection Agency has sponsored a program to construct a management plan for Galveston Bay. This would add both breadth and a long-term perspective to management. Committees were formed to address policy, public interest, and scientific issues. The entire fishing industry (fin fish and shellfish) was represented by one person, who was placed on the public interest, not scientific, committee. Local knowledge about the bay was consequently peripheralized by not including a fisher on the scientific committee. We include this case to point out that fishers are viewed as an interest (i.e., political) group rather than a group possessing valuable knowledge about natural resources. This view is evident in the stories regulators tell about fishers in Oyster Advisory Committee transcripts as well as in the deliberations on the coastal management plan.

The Texas oyster management case illustrates the need for government resource managers to expand the institutional arenas in which fishers are included. If fishers were valued for their knowledge about bay resources, they would be invited to participate in technical/scientific arenas. They also need to be included in the deliberations of agencies and as part of agencies that are not directly involved in oyster regulation: for example, agencies concerned with water quality and habitat are vital to sustaining healthy oyster stocks. Currently, these concerns are addressed by a separate agency, the Texas Water Commission, as well as by the Resource Protection Division of the TPWD. The inclusion of industry people on technical advisory committees in these areas would be a salutary first step toward empowering fishers to address the ultimate causes of declining oyster stocks.

The Texas oyster industry would be a problematic candidate for local management at this time for two reasons: (1) it is fragmented and (2) the fundamental causes of oyster depletion are not under the control of either the industry or the regulatory agency. Although they oppose viewpoints that see them as major contributors to the resources' decline, fishers are nonetheless inclined to seek regulation or let regulation stand that restricts fishing. This was illustrated by the OAC's refusal of an offer by the CFD management team to suspend all regulations regarding oysters.[17] Perhaps this refusal is indicative of a lingering mistrust of the TPWD, wherein offers to remove some restrictions were viewed as covert efforts to destroy the industry. This view was in fact expressed by a few industry interviewees. Embedded in this view is a perception of human nature that is unvoiced but in basic agreement with the CFD biologists who say that if humans are given a public resource to use with no restriction, they will "keep taking until the last one is gone." Otherwise a laissez-faire approach by the TPWD would not adversely affect the industry. Industry's refusal of any such offer may constitute a recognition of the economic equalizing effect some restrictions such as bag limits have within the industry. Whatever the reason, at this point in time, interviews with industry members revealed a contradictory stance toward regulation, one in which they decried state intervention, yet viewed the offer of nonregulation as a trap.

The second problem is, at this time, more intractable. Industry's conflicted stance toward state regulation is in fact realistic given the nature of the fragmented institutional milieu. Although the TPWD is responsible for the health of the resource, it does not have the power to stop the destruction of oyster habitat. Hence, given the limited authority of the CFD, industry input might actually be more effective in other regulatory arenas, as mentioned above. Recognition by both parties of their limited power to arrest the decline of oyster stocks using current regulatory strategies, and of how overfishing affects the resource relative to these other concerns, might engender more cooperation between them on the relatively smaller issue of overfishing, and instead redirect resources and research efforts toward better measuring the effects of freshwater retention, pollution, and dredging.

Lessons for Modern Fisheries Management

1. More effective fisheries management, as well as a better relationship between managers and fishers, would result from a redirection of managerial efforts away from controlling fishers to controlling habitat.
2. Rather than focusing on controlling fishers, managers should push at the boundaries of their jurisdiction to address pollution, salinity, dredging, and other habitat-focused issues.
3. Fishers and environmentalists should support managers in efforts to improve habitat, thus creating new alliances.
4. Industry representatives should be included on technical, scientific, and non-industry-related task forces and advisory committees.

Notes

1. For a discussion of scientific knowledge as local knowledge, see Turnbull 1992.

2. See Albrecht 1990 and Pinkerton 1989 for case studies.

3. Other studies on fisheries management that look at the roles, values, training, and bureaucratic constraints of fishery managers and policymakers include: Callon 1986, Durrenberger 1990, Marasco and Miller 1988, McCay 1988, Miller and Gale 1986, Miller 1987, Smith 1990. For studies of the relationship between university scientists, regulatory scientists, and fishers, see McCay 1988, Miller 1987, Smith 1990, and Weeks 1991.

4. Scientists in CFD formulate policy by drafting recommendations to the governor-appointed Texas Parks and Wildlife Commission, the body that has legal responsibility for setting policy.

5. TPWD has had management responsibility for oysters since the 1800s, under a variety of agency names. Before 1985, policy was set by the Texas Legislature and was statutory. The Sixty-ninth Legislature gave the department the authority to manage through regulation, removing the legislature from the day-to-day management of resources, thus giving the department a freer hand.

6. This is changing in some training programs. The Department of Wildlife and Fisheries at Texas A&M University, for example, now offers a human dimensions track. This new program was not in place at the time our interviewees were getting their degrees. Their education reflects conditions described in the 1970s articles cited.

7. We do not imply that these characterizations are not true. We are not in the position to judge the veracity of portraits managers paint of fishers and vice versa. We use the term *stories* instead of *observations* because not all of the characterizations managers made of fishers were based on their own observations.

8. See Weeks 1991 for a fuller account of the monitoring program, its history, and controversy surrounding it.

9. These criticisms directed at CFD have died down somewhat and were representative of a more adversarial relationship that existed under a previous CFD manager.

10. They feared saltwater intrusion because salt and fresh water separate into layers, with salt water being on the bottom. Thus they feared deeper channels would allow more salt water into the bays.

11. Representatives of the oyster-harvesting industry maintain that TPWD lost the suit because their monitoring data was inadequate. TPWD maintains they lost the suit because the court found that the department did not strictly adhere to the tenets of the Administrative Procedures Act when giving notice of the closure. TPWD still stands by its monitoring data.

12. See also Pinkerton, in press.

13. The notion that fish populations can be depleted is now widely accepted for most fishes. We should remember that such was not always the case. In the last century the sea was seen as limitless, and depletion was not viewed as a problem.

14. See Weeks 1991 for further discussion of the difference between regulatory and university scientists, and Weeks 1991 and Pinkerton, in press, for discussion of the relationship between knowledge of university scientists and local knowledge.

15. The differences of opinion regarding "good" science and oyster regulations are discussed in Weeks 1991.

16. See Weeks 1991 for further analysis of transcripts.

17. No transcripts for this OAC meeting were available as the meeting was held before the Department began to record OAC meetings. This story about the CFD's offer to suspend regulations was told to us both by CFD scientists and by fishers.

References

Albrecht, Daniel
　1990　Co-Management as Transaction: The Kuskokwim River Salmon Management Working Group. Master of arts thesis, McGill University, Montreal, Quebec.

Barley, Stephen R.
　1988　On Technology, Time, and Social Order: Technically Induced Change in the Temporal Organization of Radiological Work. *In* Making Time: Ethnographies of High Technology Organizations. F. Dubinskas, ed. Pp. 123–169. Philadelphia: Temple University Press.

Callon, Michael
　1986　Some Elements of a Sociology of Translation: Domestication of the Scallops and the Fishermen of St. Brieuc Bay. *In* Power, Action, and Belief: A New Sociology of Knowledge? J. Law, ed. Pp. 196–233. London: Routledge and Kegan Paul.

Chapman, Douglas
　1979　Fisheries Education as Viewed From Inside. Fisheries 4(2):18–21.

Donaldson, John
　1979　Fisheries From the State Perspective. Fisheries 4(2):24–26.

Durrenberger, E. Paul
　1990　Policy, Power, and Science: The Implementation of Turtle Excluder Device Regulations in the U.S. Gulf of Mexico Shrimp Fishery. Maritime Anthropological Studies 3(1):69–86.

Dyer, Christopher, Duane Gill, and J. Steven Picou
　1992　Social Disruption and the *Valdez* Oil Spill: Alaskan Natives in a Natural Resource Community. Sociological Spectrum 12(1):105–126.

Foster, Kevin, and John Poggie
　1992　Customary Marine Tenure Practices for Mariculture Management in Outlying Communities of Ponape. *In* Coastal Aquaculture in Developing Countries: Problems and Perspectives. R. Pollnac and P. Weeks, eds. Pp. 33–53. Kingston, R.I.: International Center for Marine Resource Development, The University of Rhode Island.

Gusfield, Joseph
　1981　The Culture of Public Problems: Drinking-Driving and the Symbolic Order. Chicago: University of Chicago Press.

Hardin, Garrett
  1968   The Tragedy of the Commons. Science 162:1234–1248.

Hester, F. Eugene
  1979   Fisheries Education From the Federal Perspective. Fisheries 4(2):16–17.

Hunter, Richard
  1984   Managerial Professionalism in State Fish and Wildlife Agencies: A Survey of Duties, Attitudes, and Needs. Fisheries 9(5):2–7.

Johannes, R. E.
  1981   Words of the Lagoon: Fishing and Marine Lore in the Palau District of Micronesia. Berkeley: University of California Press.

Marasco, R. J., and Marc Miller
  1988   The Role of Objectives in Fisheries Management. *In* Fishery Science and Management: Objectives and Limitations. W. Wooster, ed. Pp. 171–186. New York: Soringer-Verlag.

McCay, Bonnie
  1988   Muddling Through the Clam Beds: Cooperative Management of New Jersey's Hard Clam Spawner Sanctuaries. Journal of Shellfish Research 7:327–340.

Miller, Marc
  1987   Regional Fishery Management Councils and the Display of Scientific Authority. Coastal Management 15:309–318.

Miller, Marc, and Richard Gale
  1986   Professional Styles of Federal Forest and Marine Fisheries Resource Managers. Journal of Fisheries Management 6:141–148.

Pinkerton, Evelyn
  1989   Attaining Better Fisheries Management Through Co-Management: Prospects, Problems, and Propositions. *In* Co-Operative Management of Local Fisheries: New Directions for Improving Management and Community Development. E. Pinkerton, ed. Pp. 3–33. Vancouver: University of British Columbia.
  (In press) Where Do We Go From Here?: The Future of Traditional Ecological Knowledge and Resource Management in Native Communities. *In* Traditional Ecological Knowledge and Environmental Assessment. P. Booth and B. Sadler, eds. Ottawa, Ontario: Canadian Environmental Assessment Research Council.

Polunin, Nicholas V. C.
  1985   Traditional Marine Practices in Indonesia and Their Bearing on Conservation. *In* Culture and Conservation: The Human Dimension in Environmental Planning. J. McNeely and D. Pitt, eds. Pp. 155–179. London: Croom Helm.

Quast, William, et al.
  1988   Texas Oyster Fishery Management Plan Source Document. Texas Parks and Wildlife Department, Austin, Texas.

Schoning, Robert
  1984   Some Impacts of Resource Data Use in Fishery Management. North American Journal of Fisheries Management 4:1–8.

Smith, M. Estellie
  1990   Chaos in Fisheries Management. Maritime Anthropological Studies 3(2):1–13.

Turnbull, David
  1992   Local Knowledge and Comparative Scientific Traditions. Paper presented to the Society for the Social Studies of Science Conference. Gothenberg, Sweden, August 1992.

Vernon's Texas Code Annotated.
  1985   Parks and Wildlife Code 76.301, Section 1.0001-100.

Weeks, Priscilla
  1991   Managing the Reefs: Local and Scientific Strategies. Paper presented to the American Anthropological Association. Chicago, November 1992.

5

Folk Management and Conservation Ethics Among Small-Scale Fishers of Buen Hombre, Dominican Republic

Brent W. Stoffle, David B. Halmo, Richard W. Stoffle, and C. Gaye Burpee

On the north coast of the Dominican Republic, there is a small fishing village called Buen Hombre, which lies in the middle of one of the longest uninterrupted stretches of coastal mangroves in the region. Inland from the mangroves, trees and vegetation cover the hills and mountains, protecting the soil from erosion to the sea. Offshore, and extending for thirty kilometers in either direction, is a series of coral reefs. The Buen Hombre reef ecosystem is one of the most vital and biologically diverse in the Caribbean, even though it has been intensively fished for 100 years by local villagers.

Neighboring reef systems to the east of Buen Hombre are degraded due to overfishing and tourist activity. In addition, sister reefs just across the border to the west in Haiti are dead, with the longest fish taken rarely exceeding four inches (Brass 1990).

What is different about Buen Hombre? Why are its reefs healthier than those of its neighbors? There seem to be three related explanations: (1) isolation, (2) social mechanisms for adjusting the population size of the village, and (3) a strong marine conservation ethic among local fishers. The village is in a remote location, and the road connecting it to the main highway is rugged and impassable during moderate rainfall. The condition of the road has helped to prevent a substantial influx of tourists. Buen Hombre was established in 1897 by a few immigrant families from Cuba.

The population has grown since Buen Hombre's founding, but otherwise the community has developed mechanisms for reducing the resident population during times of drought and economic depression. This prevents the local population from exceeding the sustainable productivity of the coral reef, coastal mangrove swamps, and hillside farms. Finally, and perhaps most importantly, the fishers of Buen Hombre have a strong conservation ethic. The rationale of this ethic is expressed in their desire to preserve the reef's ability to supply fish for their children and for their children's children.

Today, there are forces in motion that can destroy the Buen Hombre coral reef ecosystem. Without intervention, the reef may be destroyed due to tourism development, deforestation in the nearby coastal mountains, overfishing by local fishers, and destructive fishing practices by outsiders. Social and environmental assessment studies suggest that extended drought, a depressed economy, and the rise of tourism can directly or indirectly cause the reef to die. When such phenomena are experienced worldwide, they can bring about global change in the planet's coral reef ecosystems.

The purpose of this chapter is to better describe how this community of less than 800 people, including forty-five adult fishers, preserves exhaustible resources of the fragile coral reef ecosystem and the marine animals that live on or about the coral reef. Both of these resources are vital to the existence of this small fishing and farming community. Reef destruction or the elimination of fish species could have serious adverse effects on the people of Buen Hombre. This chapter also focuses on how collaborative research with local people and government officials, combined with the use of satellite technology, can help to empower *natural resource communities* (Dyer, Gill, and Picou 1992).

The data were obtained as a result of a long-term research commitment that began in 1985 (Rubino and Stoffle 1989, 1990; R. Stoffle 1986; Stoffle and Halmo 1991, 1992; Stoffle, Halmo, and Stoffle 1991; Stoffle et al. 1993). The present analysis derives primarily from a Human Dimensions of Global Change project funded in 1990 and 1991 by the Consortium for International Earth Science Information Network (CIESIN). The purpose of the project was to understand how satellite imagery can be combined with interdisciplinary research teams to understand global change issues in coral reef ecosystems (Stoffle and Halmo 1991). The researchers used survey interviews, key expert interviews, and focus-group interviews employing satellite imagery as a basis for generating discussion. These various data sets were integrated with extensive participant observation and full access to the records of the local fishers' association.

Our studies suggested a plan for reducing the regional, national, and local stresses on the coral reefs by (1) developing an ecologically protective fish-production project using mariculture and (2) empowering the local people to help protect and co-manage the local natural resources.

Human Ecology of Fishing in Buen Hombre

A wide variety of tropical reef fish species are harvested by Buen Hombre fishers. Large groupers and red snappers are first-class fish that have a high market demand. Delicacies such as octopus, conch, and lobster are captured by spearfishers diving inside the inner reef. Coral-eating parrot fish of several sizes and varieties are either sold or consumed in local households. The size and variety of parrot fish determine whether it is classified as first- or second-class. Third-class fish species are kept primarily as subsistence fish. They bring the lowest price in the market, if and when they are sold. Occasionally, barracudas and sharks are also taken. Shark is captured for sale, and barracuda is generally kept for home consumption.

Organization of Production

Fishers in Buen Hombre rarely fish alone, preferring instead to go to the sea each day with a team of fellow fishers, called a fishing crew. By working together there is strength and security, a sharing of personal equipment and the costs of renting equipment such as boats and motors, which are not owned by most fishers. Fishing in groups also reduces the risk of returning with a catch too small for family subsistence, because the crew can be relied upon to share when needed. Fishing in groups also results in social leveling, because it assures that no fisher will become wealthier than another.

The fishing crew is the primary unit of production, but the fishers also have organized themselves into a voluntary fishers' association in order to increase the market price of their catches and to reduce spoilage. The fishers' association is composed of men who have risen through the ranks of the developmental cycle of fishing, which involves four distinct stages: (1) apprentice, (2) journeyman, (3) craftsman, and (4) "beached" (R. Stoffle 1986:95–100).

The occupation of fishing is valued positively as a way of making a living. Interview responses also illustrate the advantage of spreading risk by participating in more than one occupation, something that McGoodwin (1990:116–118) and others have described as "economic and occupational pluralism." The locally adaptive strategy of engaging in multiple occupations (fishing and farming), based on mixed production of diverse commodities (varieties of seafood and crops), serves to reduce the risk of economic failure. Perhaps an under-recognized adaptive function of occupational multiplicity is that such a system also potentially helps to reduce the risk of environmental degradation that might result from overuse of the terrestrial and marine ecozone components of coastal ecosystems.

There is a good deal of experiential learning required before one is formally identified as a fisher. Apprentice fishers typically are young boys and are not members of the fishers' association.

Entrance into the association normally occurs when novice fishers have achieved the status of journeyman.

Males are oriented to and trained in fishing at an early age by their parents. Prior to achieving status in the hierarchy of fisher classifications, the novice is expected to demonstrate certain levels of knowledge and skill (R. Stoffle 1986:95–96).

Once inducted into the association, fishers are members of a corporate group that provides services and confers access to resources. In turn, fishers take on social obligations and responsibilities for educating new members, maintaining fishing equipment, looking out for the safety and welfare of others while fishing, promoting unity, and abiding by the locally instituted informal rules of proper fisher behavior.

In 1991, six craftsman fishers were separately interviewed about fishing and conservation of the coral reef ecosystem. Each of these expert fishers was asked, "For you, what is the importance of the fishers' association?"

Simon: The association is good; we can work together if we are organized.

Dionicio: First, in unity there is strength. Second, alone a human cannot do much. Together we are looking for a project that will help us find fishing equipment — boats, motors, and cages.

Narciso: The importance of the association is that here it is like a school. We teach fishermen to protect the environment, and teach the laws of fishing and natural resources, also agriculture. It is easier to obtain help.

Bacilio: The association provides access to resources and equipment for all different types of fishing.

Eugenio: The association promotes unity. We have more strength when unified.

Tuba: The importance of the association is that we always work together; we are a group of fishermen who are united and promote the protection of the *fauna marina* [marine animals]. We don't cut coral, and we protect the turtle.

Fishers were next asked, "What are the obligations of membership in the association?"

Simon: To follow the rules of the association.

Dionicio: To carry out and fulfill the duties of the association that are required by the laws of the association. These include maintaining the equipment and obtaining fair credit.

Narciso: To maintain the order of law and keep up our equipment. To report exploitation of the mangrove and the activities of other fishermen who are prohibited from fishing here. Also we have an obligation to teach other members of the association and to share resources with its *socios* [partners].

Bacilio: To sign things out and take care of equipment, and when we get back from fishing, we have to sell fish to the association. We do this to pay for the usage of the boats and motors.

Eugenio: We have the responsibility to take care of equipment and look out for one another.

Tuba: The obligations are to try to attend meetings.

As fishers move through the hierarchical stages of their fishing careers, they tend to become more influential in the association. Individuals can aspire to the presidency (Narciso is current president) or to the position of secretary (Dionicio is current secretary) of the association. Eventually, however, fishers pass into the final stage of their careers, known as "beached" association member. Though occasionally able to fish, such members do not venture out very often and instead impart their knowledge, gained from many years of experience, to younger fishers.

All of the senior fishers we interviewed were in agreement that the productive life of a fisher lasts until the individual is between forty and fifty years of age. Fishing after this point is a physically taxing and potentially harmful occupation. Ear problems, for example, are common among fishers who spend many years and long hours repeatedly making deep dives in order to catch fish. Additionally, as lung capacity decreases with age (a problem intensified because many fishers smoke cigarettes), the amount of time they can remain submerged also decreases.

Boat and motor rental can cost a fisher and crew more than one-half (50 percent to 60 percent) of the value of the catch. If rental fees for snorkeling gear and spear guns are added, the total cost of equipment rental could total as high as three-quarters of the worth of a fisher's catch.

Traditionally, Buen Hombre fishers have employed sustainable methods of fishing that seem to derive from a conservation ethic. Interviews with local expert fishers indicate that they recognize the potential adverse effects of indiscriminate fishing practices on reef fish populations. Small fish of all classes are not targeted by fishers and are kept only rarely when captured in fish pots. Expert fishers further explain that small fish are avoided in order to allow them to grow to an appropriate size. Economically, small fish are not ideal for consumption or sale because of the low proportion of meat. Larger fish, on the other hand, provide higher returns in terms of the amount of protein-rich food compared to the amount of energy expended to catch them. Avoidance of small fish and other seafood species shows that fishers are cognizant of the ill effects that targeting these fish might have on sustaining these resources.

The enterprise of fishing entails the dual goals of providing food and income. Consequently, fishers harvest a diversified supply of seafood. Daily individual catches usually include an array of parrot fish, grouper, snapper, crab, lobster, conch, and other reef fish. The diversity of catch clearly indicates that multiple species are deliberately sought. Buen Hombre fishers employ fishing strategies for both subsistence and cash. And, just as clearly, diversifying the catch may reduce the risk of overfishing certain species.

Weather and Production Strategies

Our data suggest that local fishing strategies change based on such factors as weather conditions and stress in other sectors of the local economy (Stoffle et al. 1993). These changes can be either short-term (day, week, month) or longer-term (seasonal). Under conditions of environmental and economic stress, which are associated with drought and crop failures, respectively, Buen Hombre fishers appear to be intensifying their fishing efforts in terms of (1) length of fishing trip, (2) more intensive exploitation of certain locations along the coral reef, and (3) a concentration on capturing species that are in high demand in the market economy.

A major factor affecting fishing strategy is weather. Wind and rain play significant roles in decisions regarding whether or not one goes out to fish. Thus, in general, if the weather is favorable, the pressures to fish long hours and exert great amounts of effort are reduced. On the other hand, when weather conditions are adverse, the lack of larger boats and outboard motors mitigates against going out to the reefs to fish. Consequently, fishers may be more likely to walk along the shore to the point of the lagoon and then swim out to fishing spots well inside the inner reef. To compensate for lost subsistence and income on those days when weather conditions are not favorable, fishers may exert more effort while fishing or target certain species of seafood on subsequent days when the weather is favorable.

As is common in the tropics and subtropics, there are pronounced seasonal variations in climate. Dry and wet seasons are characterized by different weather and precipitation patterns, as well as variations in the availability and abundance of specific natural resources. Seasonal variations affect the economic welfare, health, and nutritional status of local rural populations (Chambers 1982, 1983).

In late autumn and winter, fields are prepared and planted, and the rainy season begins. While crops are ripening in the fields during this time, people must rely more heavily on purchased goods and marine resources obtained from fishing. Fishers engage in agricultural labor during the late fall and winter, working both their own fields and for wages on other farms. Fishing patterns also change, depending on the fishers. Some individuals fish more, and in more than one shift, whereas others say that hunger and lower levels of energy hamper their ability to expend a great deal of effort on fishing. One fisher suggests that sites closer to shore are used more frequently under these conditions.

There is a negative synergy between the rainy (cold) season, human health and nutritional status, and increased poverty (Chambers 1982, 1983). Ill health hampers the ability of villagers to both work in the fields and fish. Because both agricultural and marine products are sold as commodities to generate income, inability to work means that incomes are often stretched, if not exhausted, during this period. Consequently, there are frequent shortages of food and cash for buying food and medicine during the rainy season.

The six craftsman fishers we interviewed were asked to map where they would fish under two hypothetical situations entailing four scenarios (two scenarios in each situation). They marked their fishing locations on plastic overlays placed on top of a 1:50,000-scale satellite image of the area. On one overlay, the senior fishers were asked to plot the sites normally visited (1) when weather conditions were favorable or (2) when boat motors and fuel were available (see Figure 5.1). Under these circumstances, fishing sites are widely distributed. This spreads fishing pressure over a wide range of coral reef locations, with a variety of fish species being targeted.

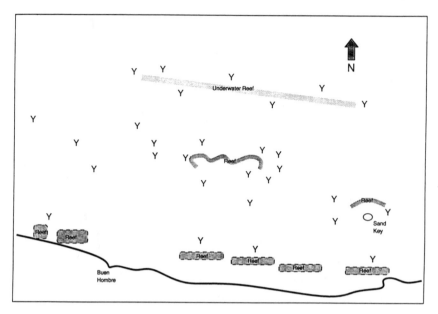

Figure 5.1 Fishing sites under ideal conditions.

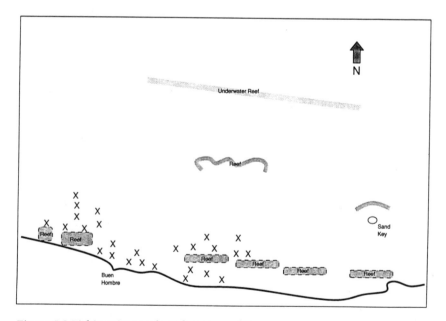

Figure 5.2 Fishing sites under adverse conditions.

On a second overlay, the six craftsman fishers we interviewed were asked to plot the sites visited when (1) weather conditions were adverse or (2) boat motors or fuel could not be obtained (see Figure 5.2). Under these conditions, the sites are concentrated on the inner coral reef. Such a strategy would result in a few locations' receiving all the fishing pressure for the entire village, with a limited set of fish species being targeted.

The second overlay correlates with the preceding discussion concerning seasonality, with particular regard to the rainy season. Because fisher-farmers are busy planting their crops and are unable to fish as much, they must buy food until the harvest can be carried out. Because crops have not yet ripened, they cannot be sold or consumed. Moreover, the nutritional value of processed foods is not as good as that of fresh foods. Colder temperatures and frequent rains also lead to an increased incidence of colds and other illnesses, especially among the young. Available cash must go for provisioning the household and investing in medication for sick family members. Consequently, access to fishing technology such as boats, motors, and gasoline may be at its lowest point during the rainy winter season. And because economic conditions are likely to be most adverse during this time of year, we can expect that the inner reef will be more intensively fished, as shown in Figure 5.2.

Our observations suggest that access to boats, motors, and fuel is more feasible during the spring and summer months, when crops have been harvested and sold and when ready cash is available. Under these conditions, the fishing locations can be expected to be both greater in number and more dispersed.

Perceived Ecosystem Threats

Like most small-scale fishers (Acheson 1988, Cordell 1989), the people of Buen Hombre perceive the coastal waters as part of their community territory. They also perceive a variety of problems, like the growing number of tourists in the area, as potential threats to the marine ecosystem. The encroachment by outsiders into the territory of the community is a problem commonly perceived by people in the village. Buen Hombre fishers, for example, expressed concern that commercial fishing fleets from nearby coastal urban centers such as Monte Cristi were depleting the fishery by using illegal net techniques that capture all sizes and varieties of fish. Concern was also expressed over competition with other small-scale fishers from neighboring coastal villages, and with a group of farmers from a distant inland village who received new fishing equipment from a church-sponsored community development project. Fishers reported to us that in 1989 the fish catch had dropped to its lowest point in twenty years and attributed this decline to outside encroachment.

The burgeoning tourism industry affects the coral reef ecozone. Even in small-scale resorts near Buen Hombre, there appears to have been an increase in harvesting coral by or for tourists for use as souvenirs. As the tourism industry continues to grow and expand beyond the boundaries of port towns, increasing numbers of tourists will intensify their search for wilderness areas, eventually subjecting the Buen Hombre coral reef ecosystem to extreme levels of disruption.

Particularly destructive for Buen Hombre has been the use of illegal *chinchorro* nets by nonlocal fishers. These are large nets up to hundreds of yards in length that extend from the surface of the water to the bottom. The nets have a catch pocket with a very fine, five-centimeter mesh size. Because of this small mesh size, the nets are capable of catching everything from shrimp to the smallest, youngest reef fish.

The way the fishers from Monte Cristi use chinchorros adds to the problem. The nets surround an area stretching from the first inner reef all the way to shore. When the nets are dropped in at the reef edge, fish are often scared out of the mangroves into the sea and then trapped as the net is brought to shore. Fish breed and take refuge in the mangroves, so this use of chinchorros can be especially destructive.

Other fishers dive for fish by using surface-based air compressors, which are connected to a diver by an air hose, enabling the diver to remain submerged for long periods to spear fish and collect other types of seafood. A large number of fish can be caught, and the use of this gear permits access to larger fish, which tend to be the breeding stock for certain species.

Senior expert fishers were asked a series of questions regarding their perceptions of the marine environment and the primary threats to its continued survival. First they were asked, "What do you think are the main threats to the ocean's environment?"

Simon: Chinchorros and compressors.

Dionicio: The sea is threatened by chinchorros and compressors. Natural things are hurricanes, storms, and earthquakes.

Narciso: For many years I have fished the sea. The most dangerous thing for the sea is chinchorro fishermen using two types of nets — *de arrastre* [drag nets] and *trasmallo* [boat-cast nets] — and compressors. Among these three things, the worst is the chinchorro de arrastre.

Bacilio: Chinchorros and compressors. These are the worst things for the sea. They take all different types of fish in their nets.

Eugenio: Chinchorros and compressors.

Tuba: Overfishing, too many fishermen, chinchorro nets, and compressors. These things affect the marine environment a great deal.

Fishers were then asked, "Do you think it is important to protect the marine ecosystem?"

Simon: Yes.

Dionicio: Yes, it is of utmost importance.

Narciso: Surely, it is our struggle to protect the sea and its animals.

Bacilio: Yes, it is very important.

Eugenio: Yes.

Tuba: Yes.

Next, we asked, "Do fishers (of Buen Hombre) do things to protect the environment?"

Simon: Yes, we fight against net fishing. We do not kill small fish.

Dionicio: Yes, we do not cut coral; we prevent the use of compressors and chinchorros. We do not cut *mangle* [mangrove], we do not take small fish and lobster, and we do not let many other fishermen in from other communities.

Narciso: Yes, we fight against chinchorro fishing and protect against cutting of mangrove and coral. We do not take small fish or lobster. We also teach our children. It is the law, but we can't maintain the enforcement of the laws we teach.

Bacilio: We do not take small fish, lobster, and [we] avoid killing pregnant lobster. We also teach our children.

Eugenio: We help those who are doing fishing by being teachers.

Tuba: Yes, we do things like not use compressors and chinchorros. We protect the fish by not killing pregnant species and small fish.

The fishers were then asked, "Do you think that fishers from other communities are a threat to the ocean environment?"

Simon: Yes, fishermen from Monte Cristi and La Varea.

Dionicio: Yes.

Narciso: Yes, certainly those from Monte Cristi and Castillo, who use chinchorros. We do not consent to the use of chinchorros here for the protection of marine life.

Bacilio: Yes, fishermen from Monte Cristi, La Vereda, and Loma Atravesada. There are also two compressor fishermen in Buen Hombre.

Eugenio: Yes, fishermen from Monte Cristi.

Tuba: Yes, because many fishermen from Monte Cristi and Loma Atravesada come in and are using compressors.

We followed up that question with, "How do fishers limit people from other villages from fishing in the territory of Buen Hombre?"

Simon: Talk with authorities. Talk with fishermen.

Dionicio: We try to work with and get help from outsiders.

Bacilio: We try and speak with government officials.

Eugenio: We try to do our part and work with government.

Tuba: We prohibit people from other places from coming into Buen Hombre territory. We work with authorities to help control the problem. We accept people coming in and fishing in our territory, but we do not want people who will come in and hurt the marine environment. They need to fish in the same way as the fishermen of Buen Hombre.

We then asked a series of questions concerning the impact of destructive technologies, such as chinchorros and compressors, on marine resources in the coral reef ecosystem. Expert fishers were asked, "Do you think there has been a change in the number of fish along the coral reef?"

Simon: Yes, there are less fish.

Dionicio: Yes, there are much less fish because of chinchorros and
              compressors.

Narciso: There is a change because people are using new equip-
              ment and overfishing the reefs.

Bacilio: Yes, because of the lack of motors, people can't go out as
              far, and for this reason they are overfishing the inner reef.

Eugenio: Yes, there were a lot more fish in the past. There are less
              fish now because of the chinchorros and the compressors.

Tuba: Yes, there has been a big change, because there are too
              many fishermen that fish every day.

We then asked, "Do you think there has been a change in the
condition of the reef?"

Simon: Yes, it is darker now. There are less fish and the bottom
              has changed. Some areas are darker and some are more
              white. Before, the reef system was clean; now it is darker.

Dionicio: No, I don't believe the reef is sick, no.

Narciso: The number of fish has changed greatly. Fifteen to
              twenty years ago, the sea was very rich with fish, but now,
              no.

Bacilio: There is more reef growing.

Eugenio: No, I haven't seen a change. The reef is in good health
              and seems to be growing.

Tuba: It does not seem sick.

These responses suggest that although the coral reef itself
remains in relatively good condition biologically, and in fact seems

to be expanding, the reef biota are perceived to be declining as a result of destructive fishing practices.

Finally, we asked each of the expert fishers, "What do you think is the future of the coral reef system?"

Simon: If we do not stop fishermen with chinchorros, we will not find fish or have resources. But if we can stop the throwing of chinchorros, production will continue.

Dionicio: If we do not control chinchorros and compressor fishing, then the system is going to die.

Narciso: I think if fishermen continue to use compressors and chinchorros, they will destroy the environment. If they kill the largest and the smallest fish, they will automatically kill whole generations of fish, right?

Bacilio: With conservation projects the reef will be protected, as well as the mangrove. Everything will be good with the reef if we continue to protect it. The reef will get sick if chinchorro fishermen continue their practice.

Eugenio: We will have to protect the reef. We think that we will have to care for the reef. We cannot do bad things to it. There will be little future if people continue to use chinchorros and compressors.

Tuba: If we do not guard and protect the reef, it will disappear. If we protect it, the reef will sustain our community.

Empowering the Local Community

The value of local knowledge and natural resource conservation practices seems little appreciated by development technicians. Small-scale fishers, however, have developed sophisticated systems of knowledge and strategies for managing the marine environments (Johannes 1981, Klee 1980, Polunin 1985). For example, if the population of a particular species of fish drops, Buen Hombre fishers will avoid catching that species for a specified period of time. Informally, local fishers have also banned the use of diving tanks because tanks provide humans with an unfair advantage over fish. At least one fisher from another village on the south side of the mountains was restricted from fishing with a tank in Buen Hombre waters by local fishers. However, because only social pressure was used to prevent a Buen Hombre farmer from fishing with a compressor, there are limits to the effectiveness of locally generated control systems.

Attempts by Buen Hombre fishers to persuade Monte Cristi fishers to fish elsewhere were unsuccessful before 1991. In fact, in 1985 Monte Cristi fishers cut the lines to Buen Hombre crab cages when they found that the lines interfered with their chinchorro nets.

In February 1991, satellite photos of the north coast reef system were shown to villagers and national-level government officials at two meetings in Buen Hombre. An interdisciplinary team of American scientists described their "sea-truthing" research using satellite photos, their inventories of marine and terrestrial flora and fauna, their social and cultural analyses, and their work with vegetative cover crops to prevent erosion. Villagers from Buen Hombre told of numerous meetings with Monte Cristi officials and fishers, at which they implored them to stop the use of chinchorros and after which they remained frustrated at being completely unsuccessful. They said they wanted these fishers barred from Buen Hombre and that they were tired of seeking resolution through peaceful means.

The Dominican officials' initial reaction was not to believe the villagers' account, saying that Buen Hombre fishers were obviously overfishing and needed to be educated. After further discussion and inspection of the project data provided by our research team, their position changed, and they realized it was the Monte Cristi officials and other fishers who needed to be educated. When Buen Hombre villagers reiterated how past communication with both Monte

Cristi officials and fishers had made it clear why the chinchorros were problematic, the Dominican officials realized that what was needed was to help the fishers of Buen Hombre enforce existing laws and to implement a program of education among the fishers of Monte Cristi.

As a result of these meetings, the vice-minister of the Department of Natural Resources planned to discuss the situation with Monte Cristi officials. The Dominican director of a natural resources conservation foundation took many notes and photos during the two Buen Hombre meetings, which he then used in a publicity campaign.

The plight of the Buen Hombre people was reported on Dominican radio and TV programs for three months. In these broadcasts the conservation foundation director noted that the reefs, mountains, and mangroves of Buen Hombre had been protected by the community. He further described a local system of fishing that avoided overburdening the marine ecosystem. He spoke of the chinchorro problem and the exploitation of marine resources by outsiders. His conclusion was that the government should intervene and take measures to stop the destruction of sea grasses and the killing of young fish populations.

Twenty-one days after the return of the team of American researchers to the United States in early March of 1991, there was a confrontation between the two groups of fishers. The two groups faced each other, sitting on overturned boats in the sand, and all of the Buen Hombre fishers were carrying knives. One of the Buen Hombre fishers, an informal village leader, suggested to his fellow fishers that the outsiders should be given a chance to avoid a confrontation. The Monte Cristi fishers were told that their nets would not be cut if they were taken out of the water.

After some discussion, the Monte Cristi fishers agreed to remove their nets and promised not to return. When a boat did return one week later, the confrontation was again mediated through discussion. The Monte Cristi fishers said that their families were starving. Buen Hombre fishers responded that their net-fishing methods were directly responsible for the degradation of fishing grounds off Monte Cristi, and that they did not want the process repeated in Buen Hombre waters. The Monte Cristi fishers left and were not seen again for four months. In that four-month period, new vegetative growth, many fish species, and many small fish reappeared on the reef. For the first time in years, great quantities of sardines were also seen by Buen Hombre fishers.

The confrontation, however, took a new twist in July 1991. A boat crewed by Monte Cristi fishers dropped its nets in Buen Hombre waters late at night, then later retrieved the nets and left before 6:00 A.M. The next day they returned in daylight with their chinchorros. They were again confronted by Buen Hombre fishers, who threatened to cut the nets. Eventually this group of fishers left. The following morning, a contingent of twenty Buen Hombre fishers went out in their boats with their knives to patrol their community waters. They would be ready for any intruders before any damage could be done. No outsiders appeared, and as of mid-October 1991 no more had returned.

Ultimately, Buen Hombre people are convinced that if pressure had not been applied to Monte Cristi officials, there would still be chinchorro problems in Buen Hombre. Villagers also said that underwater photographs taken by American researchers of chinchorros deployed by Monte Cristi fishers were a critical factor in bringing about change. These photos were evidence of illegal activity that could, presumably, be used in court if necessary. Thus, not until national government officials had proof of the accuracy of Buen Hombre villagers' accounts did they take their complaints seriously. The proof included satellite photos, inventories of plant and animal life of local and neighboring marine ecosystems, and the assertion by American researchers that Buen Hombre fishers were not the cause of the problem and that local fishing methods were in fact sustainable. The integrated tangible evidence catalyzed official action to empower the fishers of Buen Hombre to enforce sanctions against illegal activities in their community waters.

This empowerment was eventually formalized when the Dominican government issued badges to two Buen Hombre fishers who had been chosen by other fishers to patrol the communal waters. The Buen Hombre fishers now have formal legal authority to incarcerate violators and transfer them to the coast guard station on top of the mountain for further legal action.

Conclusion

The future of the coral reef ecosystem around Buen Hombre, like that of others worldwide, is in doubt. Fishers in Buen Hombre nowadays report lower fish catches and smaller fish sizes compared with those caught by the previous generation. Additionally, a dive-shop operator at a neighboring international tourist hotel states that he now must take tourists to new reefs, because during the past five years the ones close to the hotel have died. Again, fishers from distant towns are beginning to fish in the coastal waters of Buen Hombre with large chinchorro nets that are both illegal and highly destructive of fish populations. The manatee have disappeared from those areas where chinchorros have been used. The western extension of the coral reef system in Haiti has been characterized as dead, and the reefs east of Buen Hombre have been described as being fished out.

Landsat TM (terrain mapping) satellite imagery was useful in the case of Buen Hombre for visually demonstrating changing patterns of fishing behavior and distribution of site use to government officials in the Departments of Natural Resources and Fishery Resources. In response, government officials and local people can collaboratively design and implement programs of development for small-scale fisheries, as well as programs for protecting natural resources potentially threatened by human activities, by using data derived from satellite imagery and "ground-truthing" fieldwork.

Lessons for Modern Fisheries Management

1. A fishery cannot be managed effectively without the support and participation of local fishers. This is especially true in extremely isolated areas like the north coast of the Dominican Republic.
2. Local knowledge of natural history and fishers' day-to-day experiences while fishing can provide useful information for managers.
3. Small-scale, localized segments of industrial societies can learn and practice conservation under the right conditions. This is more likely to happen if the society is traditional in the sense of having a long history in the area and has a local culture that generates norms and values independent of the national culture, emphasizing a clear membership boundary for itself.
4. Despite their best intentions, even groups with a strong conservation ethic can be driven to nonsustainable behavior by natural and economic forces.
5. Local-level management can work in a complementary relationship with government rules. It is clear that local fishers who have long experience with, and a detailed knowledge of, marine environments, can serve as effective managers of the sustainable development and conservation of coral reef ecosystems.

References

Acheson, James
    1988 The Lobster Gangs of Maine. Hanover, NH: University Press of New England.

Brass, Jane L.
    1990 Haiti's Need for Mariculture? Paper presented at a USAID workshop on mariculture in developing nations, International Center for Marine Resource Development, University of Rhode Island, November.

Chambers, Robert
    1982 Health, Agriculture, and Rural Poverty: Why Seasons Matter. Journal of Development Studies 18(2):217–238.

1983 Rural Development: Putting the Last First. London: Longman Press.

Cordell, John (ed.)
1989 A Sea of Small Boats. Cambridge, MA: Cultural Survival.

Dyer, C. L., D. A. Gill, and J. S. Picou
1992 Social Disruption and the *Valdez* Oil Spill: Alaskan Natives in a Natural Resource Community. Sociological Spectrum 12(2):105–126.

Johannes, R. E.
1981 Words of the Lagoon: Fishing and Marine Lore in the Palau District, Micronesia. Berkeley: University of California Press.

Klee, Gary A. (ed.)
1980 World Systems of Traditional Resource Management. New York: John Wiley and Sons.

McGoodwin, James R.
1990 Crisis in the World's Fisheries: People, Problems, and Policies. Stanford, CA: Stanford University Press.

Pinkerton, Evelyn (ed.)
1989 Co-Operative Management of Local Fisheries: New Directions for Improving Management and Community Development. Vancouver: University of British Columbia Press.

Polunin, Nicholas
1985 Traditional Marine Practices in Indonesia and Their Bearing on Conservation. In J. McNeely and D. Pitt (eds.), Culture and Conservation: The Human Dimension in Environmental Planning. Pp. 155–179. London: Croom Helm.

Rubino, Michael C., and Richard W. Stoffle
1989 Caribbean Mithrax Crab Mariculture and Traditional Seafood Distribution. In G. T. Waugh and M. H. Goodwin (eds.), Proceedings of the Thirty-Ninth Annual Gulf and Caribbean Fisheries Institute (Hamilton, Bermuda), 1986. Pp. 134–145. Charleston, SC: Gulf and Caribbean Fisheries Institute.
1990 Who Will Control the Blue Revolution?: Economic and Social Feasibility of Caribbean Crab Mariculture. Human Organization 49(4):386–394.

Stoffle, Richard W.
1986 Caribbean Fishermen-Farmers: A Social Assessment of Smithsonian King Crab Mariculture. Ann Arbor: Institute for Social Research, University of Michigan.

Stoffle, Richard W., and David B. Halmo
1991 (Eds.) Satellite Monitoring of Coastal Marine Ecosystems: A Case From the Dominican Republic. Ann Arbor, MI: Consortium for International Earth Science Information Network (CIESIN).

    1992   The Transition to Mariculture: A Theoretical Polemic and a Caribbean
            Case. In R. Pollnac and P. Weeks (eds.), Coastal Aquaculture in Devel-
            oping Countries: Problems and Perspectives. Pp. 135–161. Kingston, RI:
            International Center for Marine Resource Development, University of
            Rhode Island.

Stoffle, Richard W., David B. Halmo, and Brent W. Stoffle
    1991   Inappropriate Management of an Appropriate Technology: A Restudy
            of Mithrax Crab Mariculture in the Dominican Republic. In J. Poggie
            and R. Pollnac (eds.), Small-Scale Fishery Development: Sociocultural
            Perspectives. Pp. 131–157. Kingston, RI: International Center for Marine
            Resource Development, University of Rhode Island.

Stoffle, Richard W., David B. Halmo, Brent W. Stoffle, Andrew L. Williams, and C.
Gaye Burpee
    (In press) An Ecosystem Approach to the Study of Coastal Areas: A Case From the
            Dominican Republic. In S. Brechin, W. Drake, and G. Ness (eds.),
            Population-Environment Dynamics. Pp. 253–282. Ann Arbor: Univer-
            sity of Michigan Press.

6

Fish as Gods and Kin
Eugene N. Anderson

In many societies, fisheries management is religiously sanctioned. Certain fish are considered sacred or are held to be kin of humans, or both. This may have some effect on management. Religious representation may make conservation more salient and deter powerful individuals from passing on costs of overuse as "externalities." It may also provide a social framework within which mutual action can occur in an otherwise individualistic or atomistic social situation.

The Question

In many societies around the world, certain fish are held sacred or otherwise regarded in a manner that anthropologists view as religious. This is part of a wider tendency to sacralize many (or all) salient environmental entities. In an earlier age, this was called animism; one locus classicus of the term is Tylor's classic *Primitive Culture* (1958), originally published in 1871. This term has now fallen out of use, perhaps mainly because Tylor associated it with "lower races." Recently, comparative work on animal sacralization

An earlier version of this chapter was delivered at the 1992 annual meeting of the Society for Applied Anthropology. It results, to a great extent, from long discussions with Evelyn Pinkerton, to whom I am most grateful. Chris Dyer and Russ McGoodwin provided extremely valuable editorial advice. Thanks also to Sharon Burton, Leslie Gottesfeld, Lynn Thomas, and others for important information and theoretical discussion, and to the people of Castle Peak Bay, Hong Kong; Kakawis and Queen Charlotte City, British Columbia; Chunhuhub, Quintana Roo, Mexico; and several other small communities I have observed.

has been left to students of religion and psychology, such as Joseph Campbell (1983), whose work is not influential in anthropology. In spite of a vast and steadily increasing literature on religious representation of the environment, there is little recent comparative study of the social and ecological grounding of this phenomenon.

Perhaps the leading exception is the work of Roy Rappaport (1984). Rappaport provided a solid framework for study and, more importantly, fearlessly proposed a theory: religion provides sanctions for ecological management rules, just as it provides sanctions for social rules. More generally, religion gives a structure to resource management. In his famous case study of the Tsembaga Maring, he held that periodic rituals involving pig slaughter kept pig populations at reasonable levels. The levels could never sink too low, because of a need for breeding stock. On the other hand, if levels became so high that gardens were menaced, a ritual reduced them. Of course, it is also possible that pig populations were deliberately allowed to build up when a ritual was planned, just as American farmers build up turkey populations before Thanksgiving. Rappaport's ideas rose well above the old, long-disproved notion that people worship what they eat. He was postulating that a society's strategies for managing the ecosystem will, in many cases, be represented or sanctioned in the religion (Rappaport 1984 and personal communication).

This chapter seeks to qualify and extend this model on the basis of case studies. I will be raising questions rather than answering them, but I have enough data to suggest that we must rephrase the debate in slightly different terms. In particular, "religion" is a problematic concept. There is a tendency in anthropology to reify or essentialize religion. The societies I discuss had no such word until they came in contact with European thought. "Management" is perhaps less problematic, but it is also a word that is not obviously grounded in visible, tangible objects.

The basic question, then, can be rephrased in more global terms: How do people convince other people to use animals — in this case, fish — in a particular way? In particular, how can people convince other people to conserve a resource, to forego using it now, even though they need it, simply because they hope to keep it for the future? This is, of course, a question with some current relevance. In spite of worldwide environmental awareness and concern, there has been little success in actually conserving fish — or forests or topsoils or water. The problem appears to be primarily one of developing sustainable management plans that will actually be accepted by users. Many excellent plans are made but not implemented. Countless laws are passed, only to be violated. Worldwide, there is failure to convince people to conserve.

There is an ancillary or corollary question addressed herein: Why are some fish worshipped? Rappaport's thought might lead us to expect that they would be the key fish in a management system, and this is a testable hypothesis. There are also testable counter-hypotheses. For instance, the worshipped fish may be those that are culturally anomalous (Douglas 1966, 1970; Anderson 1969) or those that are particularly neat and cognitively efficient symbols of cultural items (Douglas 1970). The interpretivist position, which grounds itself in abstract, rootless "meaning," should predict an arbitrary subset of local fish — a pattern not predictable from any functional consideration. Marxists might argue that particularly valuable fish — commoditized, used to signal status, or otherwise economically important — would be the candidates, or, if religion is seen as part of the validating ideology of the ruling class, that the sacred fish would be those that in some way were identified with elites. Looking at patterns of fish sacralization may thus provide insight into the Rappaport hypothesis and thus into the whole question of how people convince other people to conserve resources.

Two Cases

The Northwest Coast of North America provides a significant case in point. Here, all evidence suggests that the fish held in highest regard were the most vulnerable yet important stocks, that they were deliberately and explicitly managed, and that their social construction as revered members of human society was a conservation-related belief.

The sturgeon was held in reverence by the Katzie Salish of the lower Fraser River (Jenness 1955). According to Katzie belief, the Creator had a son and a daughter. The son became the ancestor of humanity; the daughter became the ancestor of the sturgeons. Sturgeons and humans are therefore siblings. Every sturgeon is, in a very real sense, a sister to the Katzie. This is related to the fact that the Katzie heartland — the Pitt River and Pitt Meadows area — was formerly the greatest concentration point and breeding ground for the Fraser River sturgeon population. The Katzie drew heavily on this bounty; much of their food was sturgeon flesh. As this was the flesh of their sister, reincarnating herself to sacrifice herself for her beloved family, the bones were treated with reverence and returned to the water. More to the point, no one was allowed to kill a sturgeon wantonly. Sturgeons were taken under ritual circumstances, and there was an enormously important religious taboo on taking more than was needed for subsistence (Jenness 1955).

Did Katzie conservation work, or did the Katzie simply not have enough technology to kill all the sturgeons? We shall never know, but several factors indicate the former to be correct. First, the Katzie were only one of a large number of Salish groups along the Fraser and its tributaries. These groups often cooperated with and visited each other to exploit locally abundant resources. All were superb fishers, using almost all the techniques of modern fishers except for motorized craft (see Hayden 1992, Suttles 1955). Because the sturgeons were concentrated in a very small, rather shallow, and very easily fished area, they were particularly vulnerable. They were safe in the main Fraser channel, but not in the small and shallow spawning channels of the Pitt River and Pitt Meadows. Thus, they could almost certainly have been overexploited by aboriginal technology. Moreover, the well-known Northwest Coast custom of competitive feast-giving provided a motive to wipe them out. Groups took far more than they needed, in order to assert their power and shame rivals by conspicuous consumption, which included waste. The Fraser River Salish did not have the full potlatch complex, but they had no lack of competitive feasts at which masses of food were displayed. Without a strong countervailing social rule, the sturgeon would almost certainly have been depleted.

The Katzie sturgeon cult was a local variant of a cult that was widely distributed throughout the Northwest. The Katzie also believed that one of their ancestors married a woman who turned out to be from a royal family of salmon. She taught the Katzie the correct way to fish for salmon: dispose of the bones properly and above all take no more than needed. Versions of this myth are known throughout the entire Northwest Coast area (Anderson 1987, Maud 1982). Virtually all Northwest Coast peoples believed that salmon were people who had human form under the sea and put on their fish skins only to sacrifice themselves to their friends on land. Thus, salmon skeletons, even the smallest bones, had to be returned to the water. This led to restoration of the fertility of the water. It is now known that the bodies of the spent salmon are a necessary fertilizer for the streams in which the eggs are laid. So the return of the bones provided at least some mineral nutrients. Moreover, salmon often could be taken only after ritual opening of the season, and they could not be harvested wantonly. A strong ideology of conservation still exists among Native elders, who almost uniformly insist that in traditional times there were very strong rules to the effect that people could take no more than they needed and that they should not deplete the runs (Anderson, personal research; Kirk 1986). "Need" also included the needs of the potlatch system, which could be inordinately consumptive, but the principles still hold. Today, fisheries tightly managed by Native elders are not noticeably depleting the much-reduced runs (see, for example, Morrell 1985).

Again, we lack the key data from precontact times to prove that this rule actually had an effect. However, we know that salmon were deliberately managed by at least some groups. Describing central Vancouver Island, Gilbert Sproat wrote in 1868, "It is a common practice among the few tribes whose hunters go far inland, at certain seasons, to transport the ova of the salmon in boxes filled with damp moss, from the rivers to lakes, or to other streams" (Sproat 1987:148). As Sproat was the first white man in this area, this practice may be safely assumed to be aboriginal; no outsider had come before him to introduce it. (Sailors had repeatedly visited the coast, of course, but would hardly have introduced stocking of small inland streams.) Moreover, there is proof of the effectiveness of Native conservation (and stocking) practices in the existence of runs in even the tiniest creeks. Many of these can easily be fished out by one person, using traditional technology, in a few days. This is proved, sadly, by the fact that many have suffered exactly this fate in modern times. The old beliefs have broken down, and without powerful religious sanctions there is nothing to stop poachers, Native and white. But, fortunately, in some key areas, old beliefs are strong, and fish are still conserved (Anderson, personal research; Gottesfeld 1991). This is a striking contrast to the usual behavior and belief system of non-Native fishers. Superficial observers who have charged that "the Indians" learned all their reverence for nature from the white man have clearly not spent much time talking to people in the isolated sea- and river ports.

Moreover, the complex of beliefs about the human nature of valued creatures, its insistence on respectful treatment, and its condemnation of overuse is not confined to fish. It extends to animals such as mountain goats, moose, and beaver, and to important plants such as red cedar trees. Considerable information from other parts of Canada has been amassed on traditional resource management, including the role of such beliefs (Alberta Society of Professional Biologists 1986, McCay and Acheson 1987, Sinclair 1985). For instance, the myth of the master of the game who reprimands the careless, wasteful hunter extends throughout North America, at least as far south as the Maya of the Yucatán

Peninsula (Anderson, unpublished research; Hofling 1991:136–166). There is every reason to believe that the Maya belief system is directly related to the Northwest Coast one, as the Maya myths reported by Hofling are part of a continuum of exceedingly similar stories that extends up to the Arctic (Anderson, MS. in preparation). The general plot involves a hunter who is too successful and not respectful or careful enough who is captured by his prey and taken to its house under the earth (or water). There, the chief or lord of the game informs him in no uncertain terms of how he should act.

The question of traditional resource management is reviewed at length elsewhere (Anderson 1987, Gottesfeld 1991, Hunn 1991, Swezey and Heizer 1977) and need not be further elaborated on here. Unfortunately, one cannot now ascertain the actual effect of this traditional conservation ideology. It should be noted, however, that effective conservation of fish resources via local ideology is well documented elsewhere (e.g., Berkes 1987, Johannes 1981), and conservation of plants by designating sacred groves and sacred trees is very common and effective worldwide (e.g., Alcorn 1984).

However, it is possible to establish the essential "conservationist," or at least resource-oriented, nature of the belief system by comparing it with a countercase. Cantonese fishers regard certain fish as *san yu,* "sacred fish." These include sawfish, sturgeons, whales, porpoises, and sea turtles (Anderson 1967, 1970, 1972). The reason that these are central is, quite explicitly, that they seem to bridge different categories, constituting exactly the "anomalous" and "ambiguous" creatures that provoke fear, taboo, or worship in many societies (Douglas 1966, Lévi-Strauss 1962). They seem to bridge, or violate, the boundaries between major categories. A partial exception is the sawfish. It is simply an extremely powerful fish, but its saw makes it anomalous. Occasionally the giant groupers (*Epinephalus* sp.) become so large and impressive that they have some *san,* sacred, nature. Here, the anomaly is their size — up to 200 kilograms or more — and not any unfishlike qualities.

The sturgeon is both a fish and a dragon; its weird shape and appearance and its very different anatomy give it the name *cham lung,* "deep dragon" or "sinking dragon." As a dragon, it is a potent animal of the wild waters in any case. It is rare in Hong Kong. Small ones are occasionally caught. If alive, they are thrown back, whereas if they are already dead, they are taken to the temple of the fishers' protective goddess, Ma Tsu (Tin Hau) and hung up there in a sacred place. Sawfish saws are often found with them in such locations.

The whales and porpoises are seen as *yu,* fish, but are known to have the behavior and internal anatomy of mammals. They act intelligent, and they have the flesh and inner body parts of pigs and other domestic animals. These uncanny aspects make them bridge the boundaries of land and water as well as the proper boundaries of yu. Admittedly, *yu* is broader than the English term *fish,* but mammal-like *yu* are still clearly beyond the ordinary.

All turtles, including sea turtles, occupy an anomalous status. Even the ordinary tortoise is a cosmologically important animal in Chinese thought. Moving between land and water, and sometimes burrowing or sinking into the mud, it mediates between our world and the mystic lands under the waters and under the earth. Thus it is both sacred and unclean, and many strange myths are attached to it. A sea turtle is both *yu* and tortoise and therefore doubly weird. It is never killed. Captured ones are ritually released. Buddhists hold that release of turtles and tortoises is even more praiseworthy than release of other animals. (Near Buddhist temples, there are often people selling tortoises to worshippers who want to release them for merit.) This belief has influenced the fishers, whose folk religion has major Buddhist influences.

Sturgeons, sawfish, cetaceans, and turtles are the principal *san yu.* The pattern is exactly predicted by Douglas's structural theory (1966). All water creatures seen as anomalous and powerful, and *only* those creatures, are specially protected and never killed, except by accident. There is no generalization of this view to a concept of protection or guardianship of other water life.

The very rich waters of the South China Sea appeared inexhaustible until recently and indeed were apparently inexhaustible before modern motorized fishing entered the scene. Extremely serious declines in fish stocks have occurred in the last twenty years, but this is due as much to pollution as to fishing, and the fishers know it. Therefore, a religious construction of conservation is neither warranted nor observed. On the other hand, the deep emotional involvement of the fishers in their way of life is ritually represented in these anomalous and powerful water creatures. This may suggest that, in the absence of economic motives for Rappaportian ritual, reverence for animals will still appear, but it will take a clearly social-symbolic form as predicted by structuralist theory.

The present point is to contrast the Northwest Coast belief system with that of the Hong Kong fishers. The Northwest Coast people differ in the following respects:

1. They do not have any special reverence for the anomalous fish, but rather for the most focal fish, that is, those absolutely basic for subsistence.

2. Where they do have beliefs about anomalous creatures, the belief system is absolutely different. Northern Northwest Coast peoples fear the octopus, kingfisher, and otter as part of a complex of anomalous animals associated with witchcraft. These are not held in high regard. None of the classic stories associated with conservation are told about them.

3. The classic conservation stories are told about whatever fish is particularly important for subsistence in a given area: sturgeon among the Katzie, sockeye among upriver peoples, salmon in general among many coastal groups, and, of course, goats or deer or other key game animals when one is discussing the terrestrial realm. Economic importance, not cognitive confusion, predicts the animals chosen. In both cases there is a clear pattern that must be seen in totality (Hunn 1979). This is not a crass economism; whether staple foods or uncanny anomalies, the fish are viewed with real and intense emotion. One has to see the awed reverence with which a Hong Kong fisher handles

a sturgeon, or the tender care with which a Vancouver Island Native handles a young salmon, to form a real appreciation of the power of these beliefs.

4. These fish are managed, both ritually and biologically. The myths deal quite specifically and explicitly with management strategies. In particular, the most widespread myth requires that the fish be treated with respect and that they be taken for use and never wasted.

In the case of the Northwest Coast peoples, then, hypotheses derived from Rappaport's theory are sustained.

Rappaport's Theory Extended

Decisions to manage resources on a community basis are collective decisions. They are usually decisions under risk, uncertainty, or both, for we seldom know how many fish we can take before the fishery collapses. The temptation is to guess that the maximum sustainable yield is just a little higher than we know it really should be, and then the "tragedy of the commons" is with us. Poorly shared information also leads to the more radically tragic games of the "prisoner's dilemma" family (Ostrom 1990), which typically have outcomes that are suboptimal for all players because of inability of players to share information.

One of the many good things about management at the local level is that it permits information to be shared easily; word of mouth and elementary record keeping are sufficient. This may eliminate the prisoner's dilemma condition in its pure form but leave the deciding polity still having to deal with risk and uncertainty. Moreover, it is impossible for a collectivity to be perfectly rational and perfectly democratic at the same time, because democracy involves balancing many different interests and can never have perfect information about all situations involving these interests (Elster 1983:188). It is notoriously difficult for a collectivity to be even somewhat rational and democratic. Thus, resource-management decisions should ideally be made — and perhaps can only be made — if the effective decisionmakers form a face-to-face community, probably no more than 100 to 150 people (Burton 1992). Burton points out that management can be expected to take place when a community sees that a resource is under pressure. After that, the community must then be strong enough to cope with the pressure. The stronger the community, the more pressure is manageable. Strong or multiple pressures overwhelm even a strong community. Thus, management can be seen as a summation of resource, community, and pressure ($Mg = R + C + P$) (Anderson and Burton 1992, Burton 1992).

Last of all, if a collectivity can achieve the goal of a perfect management strategy that satisfies everyone and uses the resource on a perfectly sustainable basis, it will quickly grow in wealth and probably in population, too, and thus face a new set of challenges. The younger and richer generation will have to be socialized and monitored harder than the older and poorer one was, simply because there are more players with more economic power to affect outcomes. This is all the more true when we remember that no community can be perfect. Mistakes will occur, and poaching will inevitably take place. The level of mistakes and poaching that is acceptable at Population n will almost certainly not be acceptable at Population $2n$.

Cases of institutional failure are seen in the modern United States. California's fisheries, for instance, have been decimated because of weak institutions combined with lack of concern for the future (McEvoy 1986). The United States can be seen as experimenting with the worst of all possible cases: population growth, demand growth, and technological improvement are taking place during a time when taxpayers' revolts are devastating enforcement of the law by sharply reducing management and enforcement budgets. Finally, antigovernment attitudes are leading to less and less compliance with fish-and-game laws. This last consideration introduces another set of problems. The antigovernment, broadly libertarian mood common in the United States is not specifically related to fisheries, but it has had profound effects thereon (McEvoy 1986).

The tragedy of the commons is typically legislated — or created by deliberate refusal to legislate (Anderson 1987). It is usually not a result of people's being outside any control (as Hardin [1968] argued, following Hobbes), because such people can be expected to institute a social order to manage the resource, as Hardin now more or less admits (Hardin 1991). Rather, the tragedy of the commons is typically found among people who are under the control of a strong central government that, for whatever reason, prevents those who have a real interest in a resource from organizing to preserve it (Anderson 1987, Anderson and Burton 1992, Libecap 1989). Often, as in the modern United States, the reasons have little to do with the specific resources at risk. Thus, as Libecap points out, the tragedy of the commons in U.S. fisheries today has been diligently maintained by law, in large part because of a strong ideology of free access that arose, in part, as a reaction to the British pattern of reserving game animals and fish for the gentry (Libecap 1989). Americans have been quite aware, for one or two centuries, that the British way was not what they wanted. Equality of access has been repeatedly and explicitly favored over conservation.

The emotional nature of much of the environmental dialogue in the contemporary United States (and elsewhere) serves to remind us that human decision making is typically an emotional matter. It is also subject to the notorious limits to rationality that cripple the human mind in complex situations (Elster 1983, 1989b; Kahneman, Slovic, and Tversky 1982). Emotionality is inevitable; it is impossible for humans to notice anything without at least some emotional response (Zajonc 1980), and it is difficult, if not impossible, to act in any very important way without feelings being involved. Moreover, humans naturally tend to simplify and generalize. We make the world simpler than it is, and we see it as more consistent than it is. This is the "fundamental attribution error"

(Kahneman, Slovic, and Tversky 1982:135). Also, we are bad at calculating probabilities and tend to ignore information that is not consistent with our beliefs, as well as information that is presented in a dry, unemotional manner (Nisbett and Ross 1980).

Also, humans everywhere have a very strong psychological urge or need to see the world as better and more hopeful than it is (Taylor 1989). Taylor has found that this is healthy for individuals. Unfortunately, it is catastrophic for resource management. It is the insidious rationale behind "it can't hurt to take just one," "we'll try just one more experimental fishery," "there are still lots of fish out there, we're just having a funny year," and many similar sayings familiar to fisheries managers.

It is not surprising, then, that philosophers since Plato and Confucius have warned us about emotionality and bias in decision making. Decision making under risk or uncertainty is decoupled from clear and present outcomes and is thus a prime ground for emotional distortion of the process. Also, the more long-term or wide-flung the parameters, and the less direct the feedback from actual results, the more power emotion has. The hope is always that one can avoid the costs of one's decisions, or at least pass on the externalities, that is, force someone else to pay the costs (Murphy 1967). Future generations in particular are neglected at such times. The unborn, and even the children already born, can't vote and can't sue; they are thus the ones who pay the greatest "externalized" costs. In ecology, of course, there are no externalities; someone must pay.

Decision making under uncertainty, for these and other reasons, should thus be made on a "worst reasonable case" basis. The idea is to get whatever benefits one can while being sure that the worst costs will not be devastating. One attempts to assure the "*best worst-consequence* (the maximin criterion)" (Elster 1983:187, his emphasis). This is difficult, because it is directly counter to our natural tendency to live by our positive illusions.

How can a small community deal with these problems? The first and most critical need is to maximize the amount of information —

to minimize the risk and uncertainty. Thus, the flow of information in the community becomes critical. Acquiring new information and storing and transmitting the old are thus extremely important. Such techniques have been institutionalized in some traditional societies, for example, in the fish experts of Ponape (Johannes 1981), as of course they are in modern technological society.

The second need is to have a tight social organization that has enough rationality and democracy to allow decision making but to prevent the powerful from seizing all the benefits of production while forcing the less powerful to pay all the costs. In small, relatively egalitarian communities, this may require no special institutions. Larger or less egalitarian polities, however, must have specific institutions (basically the traditional legislative, executive, and judicial) providing decision making, enforcement, and recourse.

The third need, and this is where I move into relatively less charted waters, is to provide an emotionally compelling ideology. People must be emotionally involved in the system. Otherwise they simply will not comply with its rules. Indeed, they will not feel the need to pass reasonable rules in the first place. There must be a general ideology of following the rules — sufficient to allow for whistle-blowers, citizen activists, and citizen enforcers. Also, and probably far more important, it must include a genuine love of the environment or of the resource in question, or emotional involvement in the ways of life that depend on it. Without serious, emotionally represented concern for the end states of environmental management, nothing effective will be accomplished. The modern history of conservation movements proves this. For better or worse, highly emotional movements succeed better than wise-use programs. Greenpeace succeeds in saving sea mammals on emotional grounds while sober economic and biological considerations prove inadequate to save the sardines, menhaden, cod, Mississippi River fisheries, and much more.

In a traditional small-scale society, religion serves these ends. In modern societies, a complex mixture of ideologies and demands appears to be necessary.

Obviously, from what has gone before, the less perfect the information, the more the need for emotional involvement. Moreover, information is always becoming outdated. To gather new information, people must be involved emotionally in the whole information-accumulating process (Zajonc 1980). Also, if the powerful exploit their power, the less powerful must draw on emotional bonds to achieve solidarity enabling them to mount an effective resistance (cf. Scott 1985). In practice, this usually means that there must be an emotionally compelling ideology for the less powerful (Scott's "weak") to draw upon. To put it another way, people need some kind of "cement of society" (Elster 1989) to enable them to deal with questions of this kind; appeals to individual rationality are inadequate. Institutions are also inadequate; without personal involvement, people will simply not follow laws or obey managers. For example, many countries have "paper parks" — national parks that exist on paper but not in practice — and game laws that are weakly enforced. In short, rational action and thus social institutions must be based on "emotional energy" (Collins 1991).

In small-scale traditional societies, control is very weak and rarely asserted above the village or village-group level. Where political control is weak, it is almost necessary to have some form of religious representation of community, which is what we see on the Northwest Coast. With increasing social complexity, the political world takes on a life of its own, and various political and political-economic philosophies may take on a quasi-religious role. Thus, religious involvement with the natural world may become more or less restricted to dealing with emotionally salient anomalies, as among the Hong Kong fishers. Eventually, religion may die out entirely, but a powerful emotional involvement with the natural world may reassert itself, as among many modern fishers and environmentalists. These do not worship fishing or even see management as a Christian duty. Instead, they are emotionally involved in the lifeway, and they fight for it.

It is noteworthy, however, that religion is still a powerful motive in environmentalism in modern societies. The emotional involvement probably stems from the human tendency to become emotionally involved in anything one does, but representing such matters as conservation and resource management in a communal ideology takes social negotiation from a position of power.

Conclusions and a Speculation

It is not enough to know that fish should be conserved. It is not enough to like the fish. One must convince other people to respect the fish. In fact, they must respect the fish (or the laws) so much that they will abstain from taking too many fish even if they have the chance. They must deny themselves, even when others may subsequently capture the fish.

This self-policing function is absolutely critical. No legal institution, no force of police or wardens, will be effective without this key emotional commitment. This is what the Northwest Coast myths are all about. Similarly, the many bumper stickers I observe near my home in southern California urging bass fishers to "catch and release" are apparently an effective device for persuading the fishers to forego a good dinner out of consideration for the wider fishing community. Some might see this as narrow self-interest — throw the fish back to catch it again later. However, conversations with fishers indicate that they are really choosing to do this out of a preference for the warm glow of "doing what's right" over the warm fish on the dinner plate.

The Hong Kong case shows that one should not automatically attribute all religiously based reverence for nature to conservationist causes. In many cases, quite different explanations, such as the Lévi-Strauss/Douglas theory of anomaly, fit the facts better. The Hong Kong fishers in the middle 1960s did not need to consider conservation very seriously. Only recently had their fishing pressure begun to affect the enormously rich resources of the Pearl River estuary, and they were dealing with the problem not by conserving but by either leaving the fishery or going farther offshore. They had little ideology of conservation, and their traditional religious beliefs reflect this. Rather, their religion was concerned overwhelmingly with maintaining community, curing the sick, and averting natural disasters (Anderson 1970). On land, however, where resource management was crucial and resources were very limited, the folk religion was intensely and explicitly bound up with conservation and sustainable resource use (Anderson 1970, 1987). The world's complexity is not captured by the debate between those who see traditional peoples as innately preservationist or innately destructive. (For an extreme preservationist view, see Hughes 1983 — the "simple, happy children of nature," as the cliché has it, are alive and well here. For the destructivist view, see Diamond 1992, especially pages 317–338; a more serious and thought-provoking, though still debatable, study is found in Rambo 1985.) No one in the real world fits either model, but the range among real societies stretches almost from one extreme to the other. Nor is this the only problem. For instance, it is clear that, other things being equal, the fish that generate the most emotion are precisely those that are most in demand. Thus, the pressure to overfish them is strong. Yet, less important fish may be so weakly represented ideologically that even the slightest pressure can wipe them out. If we are seeking conservation and management ideas, we have to be very careful to understand fully the societies in question.

One may speculate, pending better evidence, on the role of time. The large literature on successful management of fish and other resources in traditional societies seems to show that such resources are normally managed for sustainable yield when communities are fairly stable in identity and size over time. Too rapid change quickly invalidates plans, expectations of benefit, and institutions. Successful communities often have ways of exporting excess population, recovering from disasters, and otherwise coping with problems in a classically cybernetic way (Netting 1986). In small communities that are parts of acephalous societies, such coping is normally religiously represented. In larger societies, state and local political mechanisms may dominate. In the end, however, it is likely that people will always spontaneously come up with good and emotionally compelling schemes for managing fish, at least if left to themselves long enough; the bigger problem, and the one requiring religious representation of the community (Durkheim 1961), may be this one of maintaining a viable, stable community in spite of the vicissitudes of the world.

Lessons for Modern Fisheries Management

1. A community approach to management should be taken, with the goal being to maximize the amount of information flow in order to minimize risk and uncertainty in decision making.
2. Within natural resource communities, management design should encourage the development of a tight social organization of users based upon rationality and democracy, with that organization capable of preventing powerful outsiders from seizing control of the benefits of production.
3. Management should employ anthropologically informed strategies that provide an emotionally compelling public ideology of conservation. This ideology should support traditional management rationales of resource conservation.

References

Alberta Society of Professional Biologists
    1986   Native People and Renewable Resource Management. Edmonton: Alberta Society of Professional Biologists.

Alcorn, Janis
    1984   Huastec Mayan Ethnobotany. Austin: University of Texas Press.

Anderson, E. N.
    1969   Sacred Fish. Man 4(4):443–449.
    1970   The Floating World of Castle Peak Bay. Washington, DC: American Anthropological Association.
    1972   Essays on South China's Boat People. Taipei, Republic of China: Orient Cultural Service.
    1987   Learning From the Land Otter. Paper, Society of Ethnobiology, annual meeting.

Anderson, E. N., and Sharon Burton
    1992   Tragedies of the Commons: Malaysia. Paper, Pacific Coast Branch, American Association for the Advancement of Science, annual meeting.

Berkes, Fikret
    1987   Common-Property Resource Management and Cree Indian Fisheries in Subarctic Canada. *In* The Question of the Commons: The Culture and Ecology of Communal Resources. Bonnie McCay and James Acheson, eds. Pp. 66-91. Tucson: University of Arizona Press.

Burton, Sharon
    1992   Under What Circumstances Will a Commons Management Group Arise? Paper, Southwestern Anthropological Association, annual meeting.

Campbell, Joseph
    1983   The Way of the Animal Powers. New York: Alfred van der Marck (distributed by Harper and Row).

Collins, Randall
    1991   Emotional Energy as the Common Denominator of Rational Action. Paper, University of California, Riverside, Department of Sociology.

Diamond, Jared
    1992   The Third Chimpanzee. New York: Harper Collins.

Douglas, Mary
    1966   Purity and Danger. New York: Praeger.
    1970   Natural Symbols. London: Barrie and Rockliff.

Durkheim, Émile
    1961   The Elementary Forms of the Religious Life. Glencoe, IL: Free Press. (Originally published in French in 1912.)

Elster, Jon
    1983   Explaining Technical Change. Cambridge, England: Cambridge University Press.
    1989   The Cement of Society. Cambridge, England: Cambridge University Press.

Gottesfeld, Leslie Johnson
    1991   Gitksan and Wet'suwet'en Conservation Ethics, Concepts and Practice. Unpublished MS., Department of Anthropology, University of Alberta.

Hardin, Garrett
    1968   The Tragedy of the Commons. Science 162:1234–1248.
    1991   The Tragedy of the *Unmanaged* Commons. *In* Commons Without Trag-
            edy. Robert V. Anderson, ed. Pp. 162–185. Savage, MD: Barnes and
            Noble.

Hayden, Brian, ed.
    1992   A Complex Culture of the British Columbia Plateau. Vancouver: Univer-
            sity of British Columbia.

Hofling, Charles
    1991   Itzá Maya Texts. Salt Lake City: University of Utah Press.

Hughes, J. Donald
    1983   American Indian Ecology. El Paso: Texas Western University Press.

Hunn, Eugene
    1979   The Abominations of Leviticus Revisited. *In* Classifications in Their
            Social Context. Roy F. Ellaen and D. Reason, eds. Pp. 103–116. London:
            Academic Press.
    1991   Nch'i Wana, the Big River. Seattle: University of Washington Press.

Jenness, Diamond
    1955   The Faith of a Coast Salish Indian. Anthropology in British Columbia,
            Memoir No. 3. Victoria: British Columbia Provincial Museum.

Johannes, R. E.
    1981   Words of the Lagoon: Fishing and Marine Lore in the Palau District of
            Micronesia. Berkeley: University of California Press.

Kahneman, Daniel, Paul Slovic, and Amos Tversky, eds.
    1982   Judgement Under Uncertainty. Cambridge, England: Cambridge Uni-
            versity Press.

Kirk, Ruth
    1986   Wisdom of the Elders. Victoria: British Columbia Provincial Museum.

Lévi-Strauss, Claude
    1962   La pensée sauvage. Paris: Plon.

Libecap, Gary
    1989   Contracting for Property Rights. Cambridge, England: Cambridge Uni-
            versity Press.

Maud, Ralph
    1982   A Guide to British Columbia Indian Myth and Legend. Vancouver, BC:
            Talonbooks.

McCay, Bonnie J., and James M. Acheson, eds.
    1987   The Question of the Commons: The Culture and Ecology of Communal
            Resources. Tucson: University of Arizona Press.

McEvoy, Arthur
    1986   The Fisherman's Problem. Cambridge, England: Cambridge University Press.

Morrell, Mike
    1985   The Gitksan and Wet'suwet'en Fishery in the Skeena River System. Hazelton, Canada: Gitksan-Wet'suwet'en Tribal Council.

Murphy, Earl
    1967   Governing Nature. Chicago: Quadrangle Books.

Netting, Robert
    1986   Cultural Ecology. 2nd ed. Prospect Heights, IL: Waveland Press.

Nisbett, Richard, and Lee Ross
    1980   Human Inference. Englewood Cliffs, NJ: Prentice-Hall.

Ostrom, Elinor
    1990   Governing the Commons: The Evolution of Institutions for Collective Action. Cambridge, England: Cambridge University Press.

Rambo, A. Terry
    1985   Primitive Polluters: Semang Impact on the Malaysian Tropical Rain Forest Ecosystem. Ann Arbor: University of Michigan, Museum of Anthropology, Anthropological Papers, 76.

Rappaport, Roy
    1984   Pigs for the Ancestors. 2nd ed. New Haven, CT: Yale University Press.

Scott, James
    1985   Weapons of the Weak. New Haven, CT: Yale University Press.

Sinclair, William, ed.
    1985   Native Self-Reliance Through Resource Development. Proceedings of the International Conference "Towards Native Self-Reliance, Renewal, and Development." Vancouver, BC: Hemlock Printers for the Conference.

Sproat, Gilbert Malcolm
    1987   The Nootka. Charles Lillard, ed. Victoria, BC: Sono Nis Press. (Originally published in 1868.)

Suttles, Wayne
    1955   Katzie Ethnographic Notes. Anthropology in British Columbia, Memoir No. 2. Victoria: British Columbia Provincial Museum.

Swezey, Sean L., and Robert Heizer
    1977   Ritual Management of Salmonid Fish Resources in California. The Journal of California Anthropology 4(1):6–29.

Taylor, Shelley
    1989   Positive Illusions. New York: Basic Books.

Tylor, Edward
    1958   Religion in Primitive Culture. Paul Radin, ed. New York: Harper.

Zajonc, Robert
    1980   Feeling and Thinking: Preferences Need No Inferences. American Psychologist 35(2):151–175.

7

Local Knowledge in the Folk Management of Fisheries and Coastal Marine Environments
Kenneth Ruddle

Introduction

Over the past fifteen years, a large number of studies have revealed the deep and rich local knowledge systems on which folk management is based. Several terms have been applied to such knowledge, the main ones being *local knowledge, indigenous knowledge, traditional (ecological) knowledge, indigenous skill,* and *ethnoscience.* There are conceptual and semantic problems associated with all of them, but *local knowledge* is adopted here because it is the least problematical. Most such local knowledge has been demonstrated for agriculture; animal husbandry; forestry and agroforestry; ethnomedicine; technology; and biological, physical, and geographical phenomena (Thrupp 1988, Warren 1991).

No systemic studies of comparable scope have been made in fishing communities. However, local knowledge in fishing communities, although not usually assessed in management terms, has been described in the Pacific Basin from Malaita, Solomon Islands (Akimichi 1978); Palau, Micronesia (Johannes 1981); and

Early variants of parts of this chapter were presented at the Second International Conference of the Association for the Study of Common Property, University of Manitoba, Winnipeg, September 26–29, 1991, and at the Nordic MAB Meeting, SAS Hotel, Alta, Norway, November 5–7, 1991.

   I am indebted to Dr. R. E. Johannes, of CSIRO Marine Laboratories, Hobart, Tasmania, Australia, for his critical comments and contributions to an earlier draft of this chapter.

the Torres Strait Islands, Australia (Nietschmann 1989); as well as in Brazil (Forman 1967; Cordell 1974, 1978) and the Virgin Islands (Morrill 1967) in the New World, and in India (Raychaudhuri 1980) in the Old. Such studies have revealed that local knowledge of the environment and resources used, as well as of the society within which the resultant goods and benefits are distributed, is fundamental to the continuity of sound community-based folk management practices and to the design of new systems of sustainable resource management.

In coastal communities, bodies of local knowledge are empirically based and practically oriented. Most combine empirical information on fish behavior, marine physical environments and fish habitats, and the interactions among the components of ecosystems. This knowledge functions to ensure the sustainable utilization of aquatic resources. In some instances, explicitly conservationist objectives are evident. Local knowledge is therefore an important cultural resource that guides and sustains the operation of folk-managed systems. Knowing where, when, and how to fish, for example, governs all fishing decisions made by small-scale fishers. Further, the sets of rules that constitute folk-managed fisheries derive directly from local knowledge and concepts of the resources on which the fishery is based.

Local knowledge is of great potential value in the modern world. It can provide an important information base for local resources management, especially in the tropics, where conventionally used data are usually scarce to nonexistent, as well as providing a shortcut to pinpoint essential scientific research needs. First, however, it must be systematically collected and organized and then evaluated and scientifically verified before being blended with complementary information derived from Western-based sciences, so as to be useful for resources management.

Local knowledge is also of fundamental sociocultural importance to any society, because it provides for the maintenance of social institutions and traditional norms of behavior. Thus, just as local knowledge and its transmission shape society and culture, so, too, in reverse, culture and society shape knowledge. These are reciprocal phenomena, as is to be expected. As a result, vastly differing constructions of knowledge and processes of its transmission, as well as the social uses to which knowledge is put, occur worldwide.

Yet throughout the world local knowledge systems are changing, sometimes rapidly, owing to the pressures of Westernization, urbanization, commercialization, and monetization, and to the elitist values often engendered by nontraditional education. In a great many instances, local knowledge is thus becoming hybridized with extralocal elements.

In this chapter, I examine the social functions of folk knowledge, using brief case studies from artisanal fishing communities in Venezuela and Polynesia. Knowledge of fish behavior, marine physical environments and fish habitats, fish nomenclature, ecosystems concepts, and folk management is considered. Finally, the modern utility of local knowledge is examined.

The Social Functions of Local Knowledge[1]

The sociology of knowledge, and particularly its transmission between or among generations, is not well known (Ruddle and Chesterfield 1977, Ruddle 1991). This is extraordinary in view of the fundamental sociocultural importance of the transmission process. Similarly, little is known of the socialization and enculturation of children to community life.

Community knowledge becomes *the* given-received social world for children, and an analog of the biological-physical world with which it overlaps. Further, in the process of transmitting knowledge to a new generation, the transmitter's sense of reality is strengthened. The social world (embodied in local knowledge) becomes enlarged during transmission. It also becomes increasingly attenuated for each new generation of receivers of knowledge, to whom the rationale underlying custom, tradition, normative and actual behavior, and rules and regulations must be provided via consistent and comprehensive legitimation (Berger and Luckmann 1984).

Legitimization is extremely important, because it justifies to the individual the existence of institutions and an institutional order; not only does it justify why a person should perform certain actions and not others, but more importantly, it provides an over-arching explanation and ordering of an individual's social and cultural universe. Everything within a society occurs and becomes meaningful only within a symbolic universe. This universe provides an indigenous, theoretical, and symbolic explanation of all activities that occur; demonstrates to its own members the way in which a culture group understands itself; and makes order out of the chaos of everyday life. Such a comprehensive understanding is attained gradually during knowledge transmission. Comprehension is built on sub-bodies of knowledge such as specialized vocabularies and the behavior appropriate to them. This includes explanations of the relationships between things and behavior — such as the reasons for a food taboo — commonly explained by proverbs, maxims, legends, or folktales. For fisheries, cultural specialists provide explicit comprehensive theoretical frameworks for understanding specialized sectors of relevant knowledge and institutions (Berger and Luckman 1984).

These things in turn lead to the need for social controls to handle deviance and to ensure compliance with social norms. But because there is a need to control deviance by ensuring compliance under the threat of sanctions, the process of knowledge transmission leads logically to that of institutionalization.

Immediately relevant to local knowledge is the fact that role performance mediates appropriate parts of the overall stock of common or basic knowledge; to be a fisher naturally demands a firm grasp of the arcane knowledge of fishing. But it also requires an equally firm understanding of the general knowledge of any given society regarding how a fisher should behave within that society.

Thus, there exist stratifications of local knowledge. On the one hand are those branches of knowledge that are directly relevant to specific roles (e.g., to fishers or to farmers), and, on the other, general local social knowledge. The former are reflected in the division of labor, and therefore are accumulated faster by individuals performing their roles. Hence there emerges the specialized fisher, and within that overall category the master-fisher and traditional administrator-managers, such as elders or fisheries "magicians," whose higher degree of skills and knowledge depends on greater-than-average physical, psychological, or social attributes beyond those needed to acquire the generally shared arcane knowledge.

Role analysis is thus of paramount importance in the understanding of local knowledge and resources management. It is the main means of understanding the dynamic linkages among sectors of a society and the ways in which a collective worldview becomes manifested in the consciousness of individuals like fishers through their regular performance of routine fishing activities.

The Training of a Fisher: Examples of Knowledge Transmission

As discussed in the preceding section, the transmission of local knowledge is a critical process of education in any society. Given this, it is most surprising that the subject has been so little examined for non-Western societies. However, several of the possible forms of knowledge transmission have been documented for fishing communities. In some small-scale fishing societies, knowledge is or was transmitted via a formal apprenticeship system. For example, canoe fishers in Bahia, northeastern Brazil, pass on their knowledge through a limited number of apprenticeships that can last for as long as ten to fifteen years (Cordell 1989). On the island of Guara, in the Orinoco delta of Venezuela, children are educated in traditional food procurement tasks via a sequential, additive, and highly structured curriculum (Ruddle and Chesterfield 1977). But not all local knowledge transmission processes are as highly structured as those of Guara Island, let alone the apprenticeship system of Brazil. A striking contrast is found on Pukapuka, one of the Cook Islands of Polynesia, as analyzed by Borofsky (1987), and appears to be typical of much of Polynesia. In Polynesia, much of the corpus of local knowledge is transmitted informally, as on Rotuma (Howard 1973), for example. But, as on Pukapuka, both formal and informal patterns occur.

Guara Island, Eastern Venezuela

Apart from the monograph by Ruddle and Chesterfield (1977), there are no comprehensive studies on the transmission of local knowledge to a new generation for the Orinoco Delta. This example is abstracted from that study, undertaken on Guara Island, Venezuela. The inhabitants of Guara operate mixed household economies based mainly on shifting cultivation, some permanent agriculture, hunting, fishing, and animal husbandry. The local knowledge transmission process has the following characteristics:

1. There exist specific age divisions for task training in economic activities.
2. Different tasks are taught by adults in a similar and systematic manner.
3. Within a particular task complex (e.g., gillnetting in fisheries), individual tasks are taught in a sequence progressing from simple to complex.
4. Tasks are gender- and age-specific and are taught by members of the appropriate gender.
5. Tasks are site-specific and are taught in the types of locations where they are to be performed.
6. Fixed periods are specifically set aside for teaching.
7. Tasks are taught by particular kinsfolk, usually one of the learner's parents.

8. A form of reward or punishment is associated with certain tasks or task complexes.

This system is highly structured and systematic, with either individual or small group instruction. Emphasis is placed on learning by doing — through repeated practice over time rather than by simple observation and replication. Regardless of the complex of tasks to be taught, a teacher's first step is to familiarize the learner verbally and visually with the physical elements of the appropriate location. The entire complex is demonstrated over a period of time; proceeding additively and sequentially from simple to complicated steps, the complex is divided into individual procedures that repeat those already mastered. Finally, an entire task complex is learned, with only occasional verbal correction needed. When competent, the learner is allowed to help the teacher and to experiment and use his or her own initiative.

From infancy children become acquainted with the individual species constituting their fathers' catches, and they are constantly exposed to the gear and techniques used. The children's earliest contact with fishing is observing their mothers angling while taking a break from laundering. Both boys and girls are taught to use hook and line by an older sibling or friend and are skilled anglers by eight years of age. Girls' fishing skills are not developed beyond those needed to occasionally catch small fish for the pot, but those of boys are greatly expanded, and when a boy can handle a boat, his father starts formal training.

First, a novice fisherman goes on observational trips, watching his father closely and listening to explanations of the habits of various fish. He is allowed to assist with baiting the hooks of a trotline, a task learned earlier. After several such forays, a boy tries his hand with the trotline. When the father hooks a fish, he passes the rope to his son, showing him how to position his hands and helping him land the fish. These early attempts are made from a canoe stationed close to a levee. About 25 percent (fifteen to thirty minutes) of the fishing session is devoted to training. Generally a boy can land a fish after approximately three months of practice, comprising one or two training sessions per week. He then progresses to fishing with the trotline.

A boy is introduced to the casting net during initial training in the use of the trotline. While he observes his father casting, the net's function is explained. Then he helps to pull in a fish-laden net, keeping it close to the riverbed so fish cannot escape. After seeing full-size gear in use and gaining a feel for fishing, a lad is presented with gear proportionate to his size and physical abilities. It is with these that he becomes proficient. A father first helps his son cast the net, demonstrating step-by-step the proper throwing technique. Then the child is helped through the same sequence, starting with the folded net held in the left hand and secured to the left wrist, letting the net unfold, and finally throwing it. Although boys begin casting into the river, they often practice on dry land under the supervision of their fathers.

When a boy has demonstrated some skill in handling his small-size gear, he is allowed to fish alone or with a companion, close to the levee. But he still continues to accompany his father once or twice a week, at which time he assists with the adult gear. Boys use their own gear for a year or two until they are adjudged strong and skilled enough to participate equally in adult fishing tasks. Most boys of ten or eleven years' age are seen as adequate fishermen with both the net and trotline.

Children become familiar with net and line making at an early age. Fathers making or repairing gear often pause to tie a few loose strands of cord into a miniature net for their watching four-year-olds. These toys are used as gear for the children's small dolls. Older children receive slightly larger nets with which they can play. But not until a boy has received his own fishing net and begun to accompany his father on fishing expeditions is serious training in the maintenance of fishing gear undertaken.

A boy must maintain his own net, after first learning the technique of repair on his father's. The standard equipment for net construction and repair is two needles, a mold, and a cutting knife.

When the boy can tie the knots of the netting around the mold, he is ready to learn the use of needles. Often he sits in his father's lap and ties the netting while the man holds the net in position with his feet. The boy performs all tasks on his own, using small tools scaled to the size of his own net.

Training with the harpoon and fishing spear begins during a boy's tenth or eleventh year, when he is strong enough to handle heavy equipment and to deal with the large fish taken. Accompanying their fathers to the banks of the river, where they observe throwing techniques, boys often help pull in a catch. They are taught how to throw at convenient targets on land. A boy carries the harpoon on his right shoulder until he appears comfortable and then is shown how to throw it properly. First attempts at true harpooning, undertaken when a boy has shown a fair amount of skill throwing a makeshift harpoon, involve fish that have been brought in with the trotline gear. An eleven- or twelve-year-old is already familiar with harpooning. It takes lots of practice to become good at harpooning, and both adults and children must practice almost daily. After trying the harpoon, many prefer to use the trotline or the casting net, or even to buy their fish. It takes four to six years to really learn the art. Usually if you're good at it, it's because your father is good."

Because Guara fishers commonly bring in no more fish than can be consumed in the home and surpluses are often given to friends in the village, a boy is made aware of such reciprocal relationships during his early childhood, when he acts as messenger and carries part of his father's catch to the recipients. Unusually large catches are sold within the village, and a boy quickly learns which people do not fish and consequently are willing to buy his catch. If he is old enough to have contributed to the catch, he accompanies his father on selling rounds and thus becomes aware of likely customers. From time to time, when a surplus is large enough to warrant selling it in Tucupita, the nearby urban center, boys go along to mind the boats while their fathers haggle with potential buyers.

Fish are generally sold fresh, but salting and sun-drying for preservation are also practiced, both sexes learning the processes. Children learn to gut and scale fish during their training in domestic chores; they learn which species must have their heads removed and which can be eaten whole. With a large catch, fishers demonstrate to their children how to make slits in the fish and fill them with salt. Children place the fish they have prepared alongside those of their fathers to dry in the sun for two to five days. Usually the salting of one large catch is sufficient for a child of eight or nine to learn the process.

Pukapuka, Cook Islands, Polynesia

In Polynesia, knowledge transmission occurs within the all-pervasive context of "status rivalry" (Goldman 1970, Howard 1972, Marcus 1978, Ritchie and Ritchie 1979, Shore 1982, Borofsky 1987), or competition over status issues. On Pukupuka, such status issues relevant to local knowledge transmission are (1) social hierarchy, dependency, and deference to superiors and (2) autonomy and peer equality (Borofsky, 1987). Superior persons are deferred to by virtue of their social rank, not because they possess a superior knowledge, and, as an affirmation of their own status and worth, people challenge, qualify, or elaborate on the knowledge of others (Borofsky 1987). Further, knowledge is not always acquired or used for practical everyday purposes; an appearance of being knowledgeable and the manipulation of knowledge are used to create or enhance the status of an individual.

On Pukapuka, most knowledge is transmitted during the performance of daily tasks. This is similar to the situation on the Polynesian island of Tikopia (Firth 1936), as elsewhere in Polynesia (Ritchie and Ritchie 1979). Thus, for example, place names on a reef and the names and characteristics of reef fishes are gradually acquired as boys accompany their fathers on fishing trips. Some knowledge, however, is taught and learned for enjoyment; for example, the narration of legends provides entertainment and, over time, gradually socializes children into a group's traditions.

Verbal instruction is rare; both children and adults learn mainly by observation, followed later by imitation. Formal instruction is minimal, and questioning, especially by children, discouraged, except where it pertains to concrete situations (Borofsky 1987); and most Polynesians are visually oriented toward knowledge. Listening to the conversations of others is a second important means of acquiring knowledge. Repetition of observation, listening, and practice is the principal factor in the Pukapukan transmission of knowledge.

Learners attempt to maintain their own status vis-à-vis teachers by themselves regulating when and where they will acquire knowledge. Status is also the reason why adults do not ask questions of others, as this would imply or reveal one's own ignorance and might also cause the person questioned to either lose face or be subject to ridicule if an incorrect or inadequate answer is given. Casual, indirect conversation about a topic, however, saves face.

Challenge and indirect criticism (joking and teasing among adults) are also used as educational tools. The resultant pressure and competition provide a stimulus to learning. Thus, for the young, learning is often a humiliating and painful experience, and many people prefer to learn on their own (Levy 1973, Borofsky 1987).

The Context of Local Knowledge in Fisheries

A thorough examination of local knowledge systems must go beyond a simple description of local classifications of physical and biological environments and must analyze them within their contexts of social relationships and productive activities, among other things. This is of fundamental importance because taxonomies and other elements of local knowledge and their application have emerged from and function within specific social contexts.

The way in which a coastal marine environment is classified in cultural terms has profound implications for how it is used. However, it is important to note that taxonomies alone will not suffice to predict how, when, and where a group of fishers will behave. Reciprocally, local ways of thinking cannot be fully understood without a parallel understanding of fishers' routine behavior patterns. Thus understanding the relationships between cognition and local knowledge about fish behavior and its application in fishing activities is essential for sound fisheries management. In addition to being a key component of a fishery, the way in which local knowledge of a coastal marine environment is understood is part of a complex set of ecological relationships that involve an entire fishing community. A people's resource use patterns are products not of their physical environment and its resources per se, but of their perceptions, or culturally formed images, of the environment and its resources. Thus, to properly understand human ecological relationships, an understanding of a society's local knowledge base, and the cognitive system that underlies it, is crucial.

For example, small-scale fisheries in many parts of the Pacific Basin are governed by traditional, community-based marine tenure systems that do not distinguish between land and sea resources, which together form part of an "estate" (an estate is locally defined as a parcel of land and adjacent water controlled by one individual). Instead, both types of resources are usually conceived of as being subgroups of the same main category, with fishing grounds often classified locally, as in Yap, as "sea land" (Lingenfelter 1975). Among the Lau of Malaita, Solomon Islands, the lagoon is termed *asi namo* or *asi hara,* "sea land," and is also referred to as *raoagi ala ia,* "the cultivated field for fish" (Akimichi 1978). This is in striking contrast to many other parts of the world, where land and sea are cognitively separated and their resource use regulated by distinct management ideologies, under which they are viewed as "regulated land" and "unregulated sea" (Ruddle and Akimichi 1984).

Gender Issues in Local Knowledge

Local knowledge is "gendered" (Warren 1988:10) because men and women usually have different, and often complementary, economically productive roles and resource bases and face different sets of social constraints. Thus, some local fisheries knowledge is exclusive to the female domain. If this is not comprehended and integrated into general local knowledge, then understanding of fisheries management systems will be seriously deficient. Both consideration of logical structures of total systems of local knowledge and an awareness of gender and age roles in rural society make it self-evident that gender considerations are important in understanding local knowledge in fishing communities.

There have been few studies of the role of women in fisheries, and even fewer of their fisheries and marine-environmental knowledge. An exception is a study by Gladwin (1980) of Fante women fish traders in Ghana. His study of their knowledge of fish processing and marketing confirms that they "are highly motivated, able, and efficient" (1980:146). Ironically, Gladwin attributes the poor understanding of West African women traders to economic anthropologists, whose assertions failed "to take into account the knowledge systems these traders utilize to function in their market system" (1980:129). Yet, in fishing communities throughout the tropical world, women, and to a lesser extent children and old persons, play important roles in food production, processing, preservation, and preparation, as well as perform complementary farm work and other economic activities, and engage in trading.

Some bodies of local knowledge may have complementary male and female components, and both components are required to understand a particular aspect of fishery production. Among the Marovoans of the Solomon Islands, for example, women have an extensive knowledge of the lunar and seasonal rhythms in the occurrence of eggs and milt in many species, as they usually gut the fish brought home by men. They thus assist in the decision making regarding the use of particular fishing locations. An intricate knowledge of seasonal variations in the occurrence of crustaceans and molluscs is also possessed mainly by women, the usual collectors of these resources. The timing and locations of aggregations of land crabs, mangrove crabs, and mobile molluscs, as well as factors such as red tide that influence the edibility of molluscs, are intimately known to them. Similar knowledge, though limited to fewer species, is possessed mostly by those men who dive for commercial shells such as trochus, pearl, green snail, and tabu shells, and for bêche-de-mer (Johannes and Hviding 1987).

Significant bodies of local knowledge are overlooked when research focuses only on male heads of households or on active fishermen. Norem, Yoder, and Martin (1989) define the four main types of gender difference in local knowledge systems: women and men (1) have different knowledge about similar things, (2) have knowledge of different things, (3) have different ways of organizing knowledge, and (4) have different ways of preserving and transmitting knowledge.

Continuity and Adaptation of Local Knowledge

A coastal marine society's specialized ecological knowledge is not uncommonly an encyclopedic and complex organized body of information that has evolved through generations and is still evolving. A local knowledge system is "traditional" by virtue of its long and deep roots and its origin in a specific culture and a local ecological system, but it is not static.

Such continuity is a fundamental characteristic of any traditional system, as is its flexibility. In Third World societies, tradition is still usually unwritten and based not only on what each generation learns from the elders, but also on what that generation adds to that knowledge. For example, in many areas the use of diving technology has expanded the knowledge of fish behavior as divers encounter, observe, and stalk each fish species in its own habitat (e.g., Johannes 1981, Nietschmann 1989).

The continuity and evolution of local ecological knowledge is expressed cogently by Nietschmann (1989:65–66), writing of the Torres Strait Islanders:

> Sea knowledge and marine sciences are continually improved through firsthand experience and intellectual elaboration. From the 1860s on, Islanders expanded their understanding and description of sea conditions and sea life through new economic pursuits such as diving for pearl shell, trochus shell, bêche-de-mer . . . and crayfish. Already accomplished sea-surface naturalists, Islanders have now accumulated 120 years of direct underwater observations and descriptions of currents, biota, reefs, bottom topography, relationships between moon phase and water clarity and much, much more. *This knowledge and their long-term and continuing occupation and use of islands and the sea are what Islanders say represent their credentials of ownership* [emphasis added].

"Modern" influences do not necessarily make contemporary local knowledge less "traditional," as they are incorporated into a framework of existing knowledge. Inevitably some of the past generations' knowledge is replaced through the present's experience, but the knowledge core remains intact. This core derives from the observations and experiences of generations of fishermen and fisherwomen working in environments with which they are intimately familiar. By virtue of both this continuity and flexibility, contemporary knowledge of the coastal marine environment retains its local character.

Contact with the greater society beyond a small community generally results in the hybridizing of local knowledge with extra-local elements. But through this process local knowledge can become delegitimized entirely, with either beneficial or detrimental results, depending on context. The reaction of individuals to the impact of external knowledge varies considerably. In Costa Rica, for example, Thrupp (1988) found five overlapping categories of response among rural people: (1) increased pride in their local knowledge and methods, (2) openly expressed rejection of Western innovations and related knowledge as disruptive of local knowledge and resource management, (3) skepticism of introductions but hesitancy to express it, (4) embarrassment and shame regarding their local knowledge and techniques, and (5) idolization of the introduction and concomitant rejection of local knowledge and techniques.

Changes in the perception of environment may provide insights into the principles of marine tenure, because "the degree to which fishing rights are elaborated and enforced depends on their perceived value" (Johannes 1982:260). Thus, the perceived value of fishing privileges and changes in such valuations may be linked to alterations in the local perception of the marine environment and may modify perceptions of the general utilitarian value of different marine organisms. For instance, commercialization of a fishery or alteration of ecological or technological constraints on it may create a demand for underutilized or unused species, and new patterns of fishing regulations may emerge as a consequence, as occurred in the Trobriand Islands of Papua New Guinea (Malinowski 1935), Palau (Johannes 1981), and Tokelau (Toloa, Gillett, and Pelasio 1991).

Knowledge of marine organisms and their ecological context is primarily behavior oriented and embraces the sets of information required to locate and harvest these species. This has been demonstrated in the Solomon Islands among the Lau fishers of Malaita (Akimichi 1978) and in Marovo Lagoon (Hviding 1988), in Palau (Johannes 1981), and in Okinawa (Ruddle and Akimichi 1989), among other places. Such knowledge of fish behavior is based on personal observation of species encountered on the fishing grounds and has been accumulating through generations, each new generation verifying aspects of the previous generation's knowledge through its own experiences. Through these processes, most "inherited" items of knowledge are retained; aspects that become less relevant gradually fade away, whereas new items are added as a response to developments in the fishery. Few aspects of marine knowledge used in small-scale fisheries are not based on accumulated, empirical observation. Thus, given the capture-oriented nature of such knowledge systems, it can be reasonably anticipated that sound knowledge is verified by success in daily fishing.

Components of Local Marine Knowledge

The main component of local ecological knowledge of marine resources concerns the behavior of fish and other animals. Fish are known to behave differently in different places at different times, and such information, often combined with similar systems of knowledge about astronomical cycles, general climatological processes, sea conditions, and terrestrial resources, is a principal key to successful fishing. Whereas much of this knowledge deals with habitual aggregations of important food species, details of general fish behavior are also known, including the relationships among different reef species. Special behavioral characteristics are often embodied in fish names.

Fish Behavior

Fish behavior and that of other marine organisms is the most widespread criterion for the creation of taxonomies in non-Western traditional societies. As should be anticipated, indigenous knowledge of marine fauna behavior is particularly highly developed for species of principal economic and ritual importance. Morrill (1967:409) observed that the Cha-Cha of the Virgin Islands in the Caribbean were "conversational monomaniacs fixated on fish," for whom fish behavior was a subject of supreme interest, "taking the form of descriptions of observed behaviour, proffered explanations for the behaviour, and debate over the soundness of the observations and the conclusions reached from them," and particularly of fish feeding behavior (which extends even to species that they do not target), and to a lesser extent reproduction, territoriality, and the " 'personality' of species and individuals." Although in some societies conversation about such topics may be taboo before, during, or immediately after fishing activity, as in Satawal, Central Caroline Islands (Akimichi and Sauchomal 1982), lest it prejudice the fishing expedition, nevertheless fish behavior elicits avid discussion from most traditional fishers.

Fish behavior varies temporally according to fixed diel, monthly, and longer cycles. Thus a knowledge of the causes and characteristics of different types of fish aggregation behavior by species is fundamental to successful fishing. And the relative importance of the types of fish behavior will varying according to the types of gear employed.

Widespread is high predictability accorded to aggregations associated with reproduction, because this form of behavior correlates closely with lunar phase. Moon phase is the major indicator of predictable events on the fishing grounds, as many fishes and other marine animals form large aggregations for reproduction in known locations during certain months and moon phases.

The Cha-Cha relate reproductive behavior to both large-scale weather changes and lunar periodicity. Pelagic fishes are thought to respond to the former, as exemplified by the barracuda, which starts to breed in November, immediately after the hurricane season, whereas neritic fishes are believed to breed according to moon phase at any time of the year, except during conditions of high turbulence and turbidity during the hurricane season (Morrill 1967). Throughout the tropics, prime spots, where predictable aggregations of important food species occur, are known to local fishers. For such spots, knowledge of the timing and location of fish aggregations is capture oriented, highly prized, and often jealously guarded (Ruddle and Akimichi 1989).

Most reef fish aggregations occur at specific times of the year during a few days around the new or full moon, although some species have more extended aggregation periods. Predictable aggregations are not only marked with reference to where and on what days of the lunar month they occur, but also pinpointed by hour and by such diel change of location as might result from tidal movements, for example. This knowledge is geared specifically to the efficient and guaranteed capture of large numbers of preferred food fish. Thus, fishing for specific target species becomes predictable, as has been verified by field observations in Okinawa (Ruddle and Akimichi 1989) and the Solomon Islands (Hviding 1988), among many other places.

Marine Physical Environments and Fish Habitats

Knowledge of the marine physical environment is of major importance to both the fishing technologies employed and to the targeting of fish by species. As among the Torres Strait Islanders, physical environments and fish habitats are also denoted by often finely subdivided taxonomies that are fitted together in a coherent, interactive framework that is at once both elegantly sophisticated and eminently practical. As Nietschmann wrote of the Torres Strait Islanders (1989:66): "The sea is knowable, useable, and predictable. Marine plants and animals and sea and weather conditions are not thought of as being 'out there' in some sort of undefinable, geographic mush, undifferentiated with respect to species, season or site"; the timing, location, and conditions of the occurrence of marine phenomena are regarded as entirely predictable. As but one example of the richness and sophistication of this knowledge base, in their decision making, Torres Strait Islanders refer to a taxonomy of more than eighty distinct terms to describe different tides and tidal conditions, using a "lunar-based system that keeps track of the four daily tides, changes in height and current speed, time of occurrence and duration, seasonal shift (due to changes in earth, sun and moon orbits), water clarity, surface conditions and associated movements of fish, sea turtles and dugongs" (Nietschmann 1989:69).

Raychaudhuri (1980) describes how the fishers of Jambudwip, India, coordinate the complex variables of seabed topography, seawater conditions, and sequences of tide and ebb with fish behavior, to ensure both successful catches and their own safety at sea. In their selection of the appropriate seabed over which to conduct their activities, these fishers are "like the agriculturalists who tend to classify the soil according to its relative fertility and the types of crops grown" (1980:61). The "soil" of the seabed is classified by its capacity to support the net poles and its fertility regarding the types and quantity of fish in the waters above it. Undulations in the seabed affect patterns of water flow and the color of surface waters, which, in turn, are used as indicators of the nature of the seabed.

The color of seawater is an indicator of the presence of fish: *dab-jal,* "green coconut water," indicates an area of little importance; *gab-jal,* "mangosteen juice," indicates a general presence of fish; and *nila jal,* "blue water," and especially *juni-bhanga jal,* "water which glows like fireflies," indicates an abundance of fish. Water smell and an oily appearance to the surface are additional diagnostic characteristics. A fish shoal emits a fishy smell that enables fishers to judge its route of movement. The types of oil released vary among species and can be distinguished by master-fishers. That of the hilsa *(Hilsa ilisha),* for example, is pinkish, and, together with the large area of the shoal and characteristic disturbance of the water, distinguishes the movement of this species.

Because the movement of fish and the associated setting of nets are intimately involved with the fishers' livelihood, patterns of fish movement have been observed in minute detail. In general, it is considered that fish move in midwater, with the current and against the wind. Some species move against the current, however. Fish movements, and therefore catches, are also predicted by prior catch composition, based on the understanding of local food chains.

Fish Nomenclature

The structures of indigenous taxonomies of marine fauna differ considerably from those of Western scientific taxonomy (Akimichi 1978, Johannes 1981). Whereas the principle purposes of Western scientific taxonomy are to designate the common evolutionary origin of members of groups of closely related species and to provide a uniform international system of nomenclature, the functions of indigenous fish-naming systems are somewhat different and their structures more flexible than those of Western science. As in biological taxonomy, so in many indigenous systems: many groups of fish have a common generic name based on their anatomical similarity. But whereas biologists recognize certain fish as belonging to a single anatomical group, indigenous fishers separate them into several subgroups based on other characteristics. In Palau, for example, such classificatory characteristics include size, biting habits, fighting characteristics, habitat, association with drifting logs, distinctive habits, diet, physical peculiarity, smell or taste, and association with island customs (Johannes 1981). The Lau of Malaita, Solomon Islands, apply a complex taxonomy to living aquatic resources, based on four hierarchal levels — general, fairly specific, very specific, and varietal — that are crosscut by other ecological and behavioral categories, principally habitat type and feeding, sheltering, escape, spawning, school, and swimming-layer behavior, particularly in terms of time-space patterns and diel cycle behavior (Akimichi 1978).

Individual species of important food fishes and marine mammals are often named differently depending on the size class at different stages of their life cycle or based on other distinguishing characteristics. Different size classes are distinguished in indigenous taxonomies either because of their different habits or habitats or because of the different fishing techniques used to take them (Johannes 1981). In Palau, for example, the milkfish *(Chanos chanos)* is known as *chaol* when it is small and lives in brackish mangrove ponds and as *mesekelat* after it has grown larger and moved out to the reef to live over sandy bottoms (Johannes 1981). Also on Tobi Island in Palau, small snappers (as well as damselfish) are collectively and generically termed *richoh*. However, once the snappers have grown and moved out to deeper water, their nomenclature changes. The dull-colored species then become known by

the generic term *hatih,* "hard to find," in reference to their dull
coloration, which camouflages them in their natural habitats (Johannes 1981).

But not all societies have complex taxonomies for marine
fauna. For example, the Cha-Cha distinguish only two taxonomic
levels. The highest consists of just three groups: shark, jack, and
corail (coral), into one of which fifty-one individually named fishes
identified during fieldwork are grouped, distinguished largely by
behavioral characteristics (Morrill 1967). The category *jack,* which
includes the principal food fishes sought, comprises four Carangi-
dae species, for which there are three local terms. They are classified by the Cha-Cha according to differences in their specific body
morphology and feeding habits. The eight kinds of shark recognized by local taxonomy are classified mainly by feeding habits,
according to the type of food taken and the method of taking it.
Minor classificatory characteristics also include body morphology,
water depth in which they are mainly found, and perceived degree
of danger to humans. Because they are not sought for food and all
exhibit similar behavior patterns, the category *corail* is not further
distinguished.

Local Knowledge and Ecosystems Concepts

Although bodies of local knowledge of coastal marine systems
are not sets of ecological models based on the Western ecosystem
concept, in some societies they are nevertheless sophisticated in
terms of their understanding of local marine ecology, the behavior
of marine fauna, and the interrelationships among the organisms
involved and elements of the physical environment. An example
of such ecological thinking occurs among the Cha-Cha of the Virgin
Islands (Morrill 1967), of whom it has been said that "their analysis
of any item of fish behavior would be accepted by zoologists or
psychologists as appropriately restricted and rigorous in method of
observation and inference, though it might be incorrect" (Morrill
1967:410). In contrast to relatively simple taxonomies for marine

fauna, the Cha-Cha possess a detailed and extensive knowledge of marine ecology and fish behavior.

The Cha-Chas' sophisticated knowledge of coral reef ecology and the behavior of marine organisms is well exemplified by their logic regarding the etiology of *ciguatera* fish poisoning. Although such varied causes as manicheel berries, jellyfish, marine worms, molluscs, zooanthellae in corals, dinoflaggelates, copper from ship bottoms, and algae have been suggested, Morrill's most knowledge-able informant on the subject claimed that ciguatera occurs when reef fish eat a particular algae, which he claimed to have identified (Morrill 1967:413):

> He reasons that it is impossible that manicheel berries are the cause because highly territorial fish which prove toxic are taken in areas where there are no manicheel. Marine worms are unlikely because some fish which are toxic never eat marine worms or other fish which do eat them. Copper makes some sense as a source because ciguatera is common around wrecks, but not all wrecks have copper bottoms, and some ciguatera is found far from any known wrecks. Algae are the most probable cause because not many species of fish eat algae, only a few of these are taken as food, and these few are heavily implicated.

Marine ecological reasoning underlies the Cha-Chas' explanation of the geographical variation in the occurrence of ciguatera. Ciguatera is rare on the south and west of the island and in sheltered bays. Toxic fish taken from such areas are always piscivorous species that have presumably fed on algae-eating fish and then moved. A high incidence of ciguatera occurs where reef ecology has been disrupted by either physical damage or pollution (includ-ing salinity changes), such as near recent wrecks, former anchor-ages, stream mouths, or where explosives have recently been detonated. The explanation rests on the observation that reef algae are early succession organisms on such disturbed reefs and that the fish grazing these organisms become poisonous. These observations concur with the scientific evidence.

Elsewhere the thinking may be less sophisticated, long-term ecological implications of various types of resource use not always perceived, and target species seldom seen as integral parts of a general marine ecosystem. In some places, as in many parts of the South Pacific, this might be attributable to the many decades of strong mission influence, which has devalued the local knowledge base by insisting that God is omnipotent and has all the answers.

Although ideas of general ecosystems may not be common in local knowledge, a rudimentary awareness of various marine ecological linkages is commonplace in small-scale fishing communities. General resource depletion, manifested in either increasing scarcity or decreasing mean size of individuals, is widely commented on as being bad but inevitable.[2] The traditional idea of a rest period for stock recovery is a resource-management practice widespread in the Pacific. In many parts of the world, certain fishing methods, especially dynamite fishing, are perceived as destructive to the marine environment and the resource base. Experienced fishers commonly observe that dynamiting not only kills fish indiscriminately, but also destroys the coral habitat and therefore precludes future good fishing. Local legitimation of the widespread taboos on the use of explosives demonstrates a clear perception of causal linkages.

Wide-scale linkages might also be perceived among ecosystems. In Marovo Lagoon, Solomon Islands, for example, such an understanding extends to an appreciation of the possible wider consequences of large-scale commercial logging, mining, and fishing. Logging operations and the possibility of upland mining led locally to an increased awareness of the problem of erosion-induced lagoonal sedimentation, with its possible consequences for fishing (Hviding 1988). Commercial fishery operations have also elicited pseudo-ecologically informed responses. In Marovo the idea of food chains exists, as expressed by the awareness that bait fishing by the tuna industry depletes the prey of important food fish. Tuna bait boats have been operating there for several years, and some fishers and leaders are increasingly opposed to these activities, arguing against bait fishing, based on what they consider a "universal fact," that small fish are eaten by bigger fish, which in turn are used as human food. If the bait species are overfished, it is reasoned that food for the larger fish will become scarce (Hviding 1988). But the interactions between bait-fish stocks and other reef and lagoon fish populations are complex and not-well-understood phenomena (Evans and Nichols n.d., Baines 1985), and it is likely that these "ecological" perceptions of the Marovoans are unfounded.

In Fiji, similarly, it is widely thought by the public in many parts of the country that bait fishing has a deleterious impact on inshore fisheries, causing a scarcity of reef fish, and should be prohibited. But the perspective of the Fisheries Department, based on an examination of research data from several South Pacific nations and those from Fijian waters, is that bait fish are small, short-lived oceanic pelagics not closely associated with any particular reef. Rather, they seasonally aggregate into large nocturnal feeding schools in lagoons. Bait fish form part of the food chain for larger fishes, including food fish important for humans. It is a logical conclusion that overfishing of bait fish would reduce the numbers of the species that feed on them, and this may be why so many people are worried about the effects of the bait fishery. However, any overfishing will reduce stocks. If the number of predator fishes is reduced by fishing, then bait fish numbers will actually increase in response. According to the Fisheries Department, it is almost certain that the stocks of bait fish in Fijian waters is now as high, or higher, than at any previous time, owing to the increased fishing pressure on coastal fisheries and predator fishes. The impact of the amount of bait fishing by pole-and-line boats on this large stock of small, highly mobile, and rapidly renewable fishes is negligible; it is one of the least ecologically damaging fisheries in Fiji today and, in some areas, may even help balance the ecology of the food chain following overfishing of other species.

Conservation

Key questions in the analysis of local knowledge and resource-management practices concern the perception of resources' being finite and depletable and whether a conservation ethic exists. The term *conservation ethic* is defined here "as an awareness of people's ability to deplete or otherwise damage natural resources, coupled with a commitment to reduce or eliminate the consequences" (Johannes and Ruddle n.d.).[3]

Such questions are important to the debate on traditional community-based marine resource-management systems, for example. Some authors give preeminence to considering the systems as "marine conservation practices" (e.g., Johannes 1982). Others, however, have suggested that ecological conservation, in those instances where it can be empirically documented, is essentially an unintended outcome of a system of customary law based mainly on political concerns. Polunin (1984) takes such a stand in his examination of available evidence on such systems in Indonesia and New Guinea, contending that "sea tenure" in certain cases appears to be ecologically dysfunctional. It is important, therefore, not to assume a priori that traditional management systems are intentionally conservationist. Rather, local rationale and possible conservational functions must be looked for in each case.

Uncommon is the case of ancient India, for example. There the rich and ancient Hindu knowledge of fish, which was first documented during the period 600–300 B.C. (Hora 1935, 1948, 1952, 1953), was strongly associated with both explicit and implicit regulations for fish management and conservation (Hora 1948, 1953). In the regulations for the "Superintendent of Slaughter Houses" (Book II, Ch. 24, of the *Arthasastra* of Kautilya, the earliest-known dated work of the ancient Hindus, written in the period 321–300 B.C.), punishment and monetary fines were specified for persons who entrapped, killed, or molested fish (Hora 1948), and the *Visnu Dharmasutra* states that even the unintentional killing of fish must be atoned for by three days of fasting, "suitable penances," or fines (Hora 1953).

There are few data to demonstrate quantitatively that any local society is practicing resource conservation. The evidence concerning conservation by traditional fishers is mixed. Several types of marine resource conservation measures may have been traditionally and consciously employed by Oceanian communities to conserve stocks and ensure sustained yields. Among these were the live storage or freeing of surplus fish caught during spawning migrations, the use of closed seasons (particularly during spawning), the placing of taboos on fishing areas, the reservation of particular areas for fishing during bad weather, size restrictions (although this was uncommon in Oceania), and, in recent times, gear restrictions (Johannes 1978, 1981, 1982). Others, many of which were related to traditional religious beliefs, also may sometimes have functioned coincidentally as conservation devices.

If community-based traditional marine resource-management systems were originally designed as a conservation measure, admittedly an unprovable assumption, they would have been the most widespread conservation measure employed throughout the Pacific Basin. Widespread in Oceania is the use of closed seasons, which are determined by local knowledge about the spawning periods of key fish species and prohibit the capture of certain species during such periods. Other types of customary fishing regulations exist, often based on nonecological rationales such as religious taboos. These appear to have similar conservational implications (Johannes 1978).

Such practices are not static, and some of the new regulations that village communities devise to cope with changing fishing technology are explicitly conservationist. There are numerous examples in the South Pacific (for that of Tokelau, see below).

But not all Pacific island cultures possess a marine conservation ethic. Johannes and MacFarlane (1991) found no evidence for it among Torres Strait Islanders, for example, and they speculate that a conservation ethic is less likely to develop among people whose environmental limits are functionally unlimited (see also Chapman 1987). Torres Strait Islanders, unlike many Pacific peoples, have always had access to marine resources greatly in excess of their needs or abilities to harvest. Other peoples in the region who appear traditionally to have perceived no relationship between their fishing pressure and marine resource availability include the Ponam of Manus, Papua New Guinea (Carrier 1987). But where they exist, traditional conservation ethics and management practices provide a locally sanctioned code of behavior that can be harnessed to further the objectives of modern marine resource management (cf. McGoodwin 1984).

A concern about fisheries conservation has been expressed by villagers in widely scattered parts of Papua New Guinea. On Yuo Island of East Sepik Province, for example, fish stocks are perceived to be declining as a result of overfishing. Shellfish (not specified) too small for consumption are removed from their natural habitat and reseeded near villages, for use in inclement weather (Kainang 1984). Similarly, because of observed depletion of stocks by the villagers, giant clams are relocated in specific areas in the village of Bwaiyowa, Fergusson Island, Milne Bay Province (Yamelu 1984). In Maipenairu Village, on the northern Gulf of Papua, small fin fish and crabs were formerly returned to the sea (Frusher and Subam 1984). But no other village reported conservation measures occurring either at present or in the past.

There is concern about dwindling reef fish populations on Satawal Island in the Central Caroline Islands of Micronesia. There, fishing of one reef section is prohibited by taboo in order to provide a breeding ground to supply the rest of the reef. The area was fished on authorized days only, which amounted to just five during a three-year period (McCoy 1974). However, because unlike the reef fish, the behavior of turtles is less well known to the Satawalese, conservation measures for them are more difficult, as their regular return has engendered a faith among the islanders that they will always return (McCoy 1974).

Three types of traditional marine conservation measures have been distinguished for Tokelau, in the South Pacific, where there has been no traditional tenure system. These are specific measures, indirect measures, and the perfection of nondestructive fishing techniques (Toloa, Gillett, and Pelasio 1991). The *lafu,* or fishing taboo, is the most explicit conservation measure. It is invoked by the Council of Elders and results in a ban on all fishing in specific areas on the main reef. A lafu may be announced to permit stock recovery or to build up supplies in anticipation of a future need, such as for a festival. Other such measures are the return to the sea of undersized fish and the ban on fishing using bêche-de-mer toxins, which are known to damage corals and have long-term negative repercussions on fisheries. And, although of doubtful biological validity, the obligation to tow harvested clams around the reef to release eggs demonstrates an awareness of the need for conservation for sustained management.

The conscious perfection of nondestructive fishing techniques and skills has a conservational spin-off. Such skills are transmitted during the many years of training of a *tautai* (specialist), who uses only the "proper" fishing techniques, rather than those that necessarily give the best results. An example, based on a detailed knowledge of octopus behavior, is the preparation and use of an octopus stick to extract the animal, which obviates the need for the destructive crushing of the coral or the use of poison (Toloa, Gillett, and Pelasio 1991).

But some caveats are in order, because too much should not be made of such observations prior to scientific verification. As Johannes (1989:7) observes succinctly: "The romantic and uncritical espousal of traditional environmental knowledge and management is an extreme almost as unfortunate as that of dismissing it. Traditional peoples have not lived in some preternatural state of harmony with nature. Some of their abuses of natural resources have been, and remain, substantial."

Some claims regarding the environmental wisdom of traditional peoples are naive. For example, tabooing sacred fish species has often been described as a traditional conservation measure. But this cannot be assumed without closer investigation, because many such traditional management practices were not developed explicitly for conservation. More likely the original intent of many was resource allocation, and the rationale for this is found in local religious and supernatural beliefs, realms far removed from conservation. Further, when the taking of one locally abundant species is prohibited, fishing pressure may be increased on other, more vulnerable species. Inevitably, excessive claims for traditional environmental wisdom have provoked the equally extreme backlash that traditional environmental practices were often basically unsound. But in reality, wise and unwise environmental practices often coexist within a society (Johannes 1989).

Further, it is fundamentally incorrect to ascribe directly Western concepts of conservation and human-environmental relationships to non-Western cultures. Without detailed on-site verification, it is ethnocentric to assume that either the intent or coincidence of exclusive property rights, for example, was conservationist. In many instances in Indonesia and Papua New Guinea, for example, Polunin (1984) asserts that systems of sea tenure are a response to conflict occasioned by resource scarcity. However, this assertion, too, is oversimplified, disregarding, as it does, the distinction between resource conservation as an original intention and as an implication of traditional governance (Hviding 1988).

The relationship between exclusive tenure systems and resource conservation is further complicated by the relationship between humans and animals, respect for the latter not necessarily resulting in their conservation (Brightman 1987); present ecological situations and future environmental change (Townsend and Wilson 1987); and social norms regarding the use of resources (Carrier 1987). For example, Berkes (1987) observes that the Cree Indians of Canada regard the assumption that animals can be manipulated to ensure future productivity as an example of Western arrogance. Similarly, Carrier (1987) observes that Ponam islanders of Papua New Guinea regard animals as active agents and humans as passive. Thus, the latter can do nothing to ensure the future productivity of the former. Accordingly, among Ponam islanders exclusive fishing rights combined with traditional ecological knowledge are not perceived as functioning as a conservation device; rather, they serve to fulfil the social obligation to be generous with marine products.

Nevertheless, because the resource bases governed by many common-property regimes appear to have been sustained over many social generations, with the social objective of sustained self-sufficiency reinforced by sanctions against individual accumulation to the detriment of the community, they might logically be seen as conservationist in intent. However, this has often led to conflict with state resource law, and it clearly changes in a great many instances under the tensions introduced by development programs. In many cases, customary systems have either disappeared or become greatly hybridized under such pressures, whereas in others, they have proven remarkably resilient to progressive adaption, as in the case of Japan (Ruddle 1987).

Traditional conservation practices are increasingly beset by pressures in the modern world. These pressures include the integration of local economies in the cash and export spheres, the introduction of modern fishing gear, changes in educational systems, and the erosion of traditional authority, all of which strain the system in Tokelau, for example (Toloa, Gillett, and Pelasio 1991). Monetization has eroded the respect and thus the authority of the Council of Elders because they are now salaried and have ventured into nontraditional areas, like budgeting. It also has enabled islanders to pay relatively painless cash fines for transgressing fishing regulations. Whereas the isolation of Tokelau led to the use of marine resources exclusively to satisfy local needs, improvement in transportation links with Western Samoa has made possible overseas marketing. The export demand for giant clams (*Tridacna* sp.), in particular, has led to overharvesting and a marked decline in the resource. Further, the introduction of diving goggles, gill nets, and spear guns has also led to increased harvesting pressure.

Traditional fishing skills have declined with economic and educational changes. This is manifested in increased pressure on easy-to-catch species, like parrot fish (Scaridae), and a concomitant reduction of pressure on stocks such as giant maori wrasse or eels, the catching of which requires either special knowledge or intense physical effort.

Summary of the Main Characteristics of Local Knowledge of Coastal Marine Environments

From the above discussion, it is evident there exist several characteristics common to corpuses of local knowledge of coastal marine environments and resources in many widely separated parts of the world:

1. They are based on long-term, empirical, local observation that is adapted specifically to local conditions, embraces local variation, and is often extremely detailed.
2. They are practical and behavior oriented, focusing on important resource types and species.
3. They are structured, which makes them somewhat compatible with Western biological and ecological concepts.
4. They are often dynamic systems capable of incorporating an awareness of ecological perturbations and of merging this awareness with an indigenous core of knowledge.

The Modern Importance of Local Knowledge

The contemporary importance of local knowledge may be summarized as (1) inherent academic interest, (2) practical usefulness, and (3) an instrument of empowerment. The inherent academic interest is self-evident and will not be discussed further.

Practical Usefulness

Owing largely to disparagement, the practical usefulness of local knowledge is rarely exploited. Although it would seem an obvious thing to do, local knowledge of resources and environments is rarely used to assist the design of development projects or management systems. The tendency to disparage local knowledge is not new, although the reasons for disparagement have changed. For example, in New Caledonia the processes of discrediting localized knowledge of resources has been going on for generations among European colonists, administrators, educators, and missionaries (Dahl 1989). There, at the beginning of the twentieth century, Lambert (1900; cited in Dahl 1989:45–46) refused to record all the "superstitious ceremonies related to fishing. It is sufficient to say: pity our poor natives, may we appreciate and encourage the apostolic work, which is alone capable of dispelling such darkness." Although such an attitude culminated with Western colonialism and social science of the nineteenth century (Warren 1989), it was firmly rooted in the works of historians and natural scientists of the seventeenth and eighteenth centuries (Slikkerveer 1989).

Interrelated economic, ideological, and institutional factors still combined to perpetuate the marginalization and neglect of local knowledge and, therefore, of participatory approaches to development and management. Principal among these factors are following:

1. The bias of elite professionalism, as a result of which local knowledge lacks legitimacy in mainstream thought, which regards objective Western science as superior. Western and Western-trained scientists generally reject local knowledge, which they either cannot or will not understand, which does not fit into their formal models, and which challenges conventional theories. Local knowledge is still widely belittled at best, and projects that attempt to make use of it are frequently viewed as unscientific and therefore unacceptable. Such attitudes remain deeply embedded both in individuals and institutions, such that

persons wishing to pursue unconventional projects and research often face ridicule and, occasionally, job loss.

2. Related is the belief that empirical methodologies in laboratory settings are the only correct procedure. This top-down approach aims to uplift rural societies via standardized technological transfer. Such an approach is upheld and promoted by the organization and incentive structure of research institutions and professions and the extension services that implement their findings. Innovative approaches are thereby discouraged.

3. The private sector invariably reinforces that approach, because its continuing profits are predicated on the transfer of technology.

4. Weaknesses (now overcome) in the early attempts at participatory development approaches — especially in farming systems research — reinforced the conventional skepticism of individual scientists, institutions, and donor agencies. This has been compounded by the difficulty of showing, using conventional criteria, quantifiable results to demonstrate success and cost-effectiveness in participatory approaches.

5. Promotion of local knowledge and participatory development is viewed by some central governments as organizing the rural poor and is therefore seen as subversive.

The denigration of local knowledge as backward, inefficient, inferior, and founded on myth and ignorance has recently begun to change on the basis of evidence from numerous studies that local knowledge often has rational bases. Many local practices are logical, sophisticated, and often still-evolving adaptations to risk, based on generations of empirical experience and arranged according to principles, philosophies, and institutions that are radically different from those prevailing in Western scientific circles, and thus all but incomprehensible to them.

As a backlash to decades of denigration, there has been a tendency by some researchers to idealize, romanticize, and attribute superior capacities to indigenous communities. This, too, is unhelpful, misleading, and inappropriate. The types, sophistication, and distribution of local knowledge throughout the tropics are immensely varied, as are the capacities of individuals within societies to both comprehend and utilize their own group's knowledge base. Not all inhabitants of all fishing villages are sophisticated possessors and users of arcane knowledge. Further, historically once-well-adapted and effective knowledge quickly becomes inappropriate when external factors cause massive and rapid change in the local social, biological, or physical environment. It is important to remember that some fishers are fallible, because they, too, can incorrectly interpret natural phenomena, as when poor fishing is attributed to sorcery or supernatural sanctions or when local interpretations of phenomena conflict with established facts. But a hasty dismissal of erroneous interpretations "risks overlooking the value of the empirical knowledge underlying it" (Johannes 1989:7).

Local Knowledge and Empowerment

Local knowledge can be understood as a system of power and thus can provide a basis for the empowerment of communities to undertake folk management. This is particularly important in tropical, multispecies fisheries and their environments, for which the scientific knowledge is still relatively poor. The practical, behavior-oriented corpuses of local knowledge, which focus on the most important species and their habitats, are of immediate value to fisheries planning, monitoring, and management and to the enforcement of regulations.

Knowledge concerning the timing, location, and behavior of spawning aggregations of reef and lagoon fishes provides an illustration. Populations of individual such fish are usually thinly and unevenly distributed, making censusing (an important tool of the manager of any wild animal population) almost impossible. But when these fish spawn, they characteristically form large, often docile aggregations at very specific locations during certain seasons, often during particular lunar phases. Knowledge of the timing and location of such aggregations is of obvious value to fishers, and collectively, the fishers of Oceania, for example, know far more about the local timing and locations of these aggregations than do marine biologists (e.g., Johannes 1981, 1989). Because they occur at predictable times and places, spawning aggregations provide excellent opportunities to monitor stocks. In addition, they provide a useful focus for management because exceptionally large catches are often made from them.

Such knowledge extends beyond biology to an intimate spatial familiarity with the physical environment, including local currents and other such phenomena (e.g., Johannes 1981). It also often includes a finely detailed mental mapping system (Nietschmann 1985, 1989; Ruddle and Akimichi 1989).

In addition to their specific fisheries knowledge, local fishers often know much about a wide range of other significant biological events. This could be important in siting and managing protected areas, for example, especially where recorded scientific knowledge is inadequate (Johannes and Ruddle n.d.). This is especially important when trying to establish reserves where people traditionally depend heavily on the local natural resources. Management involving government personnel and local people working together helps strike a balance between the former's desire for self-determination and the latter's need for assurance that renewable natural resources are well managed, particularly migratory stocks that traverse several locally managed areas. Because there is no effective traditional mechanism for dealing with this problem, fishers could clearly benefit from, and perhaps even welcome, outside control over the relevant fishing practices.

Lest it be overlooked, it is important to note that local knowledge is of continuing importance to present-day subsistence fishing over most of the world. Acquiring, using, and transmitting such knowledge is still extremely relevant to livelihoods in many areas, particularly where marine resources are still relatively abundant.

Conclusion: Some Requirements for Applying Local Knowledge in Modern Management

Clearly, local knowledge in coastal marine societies is rich and varied, and, equally clearly, the potential for applying it to the management of tropical coastal marine resources is substantial. But first it has to be collected. Because research projects to record and evaluate local knowledge are rare in the Third World, although their numbers are growing, there is an urgent need to stimulate low-cost data gathering. In this, high school and university students have been used with success, as in Papua New Guinea (Quinn, Kojis, and Warphela 1984).

Once collected, this data must be verified and also blended with more technical forms of biological research, like population dynamics, population genetics, physiology, and microbiology, among others, before it can be put to best use. And linguistic analysis is sometimes required to determine the exact definitions of local taxonomies in order to ensure that fishers and scientists are talking about the same species. This, it must be admitted, is an awesome undertaking. But traditional fishers and other resource users can often guide and help focus researchers' questions and search for answers.

But the evaluation of local knowledge is not easy. First, objective criteria must be designed to evaluate it and to discriminate biological truths from a mass of myth. Such criteria need adapting to each distinct cultural group. Under ideal circumstances, local knowledge of natural resources should be recorded and evaluated by persons with a background in ecology and resources management who are versed in the skills of interpreting among cultures. In this, it is imperative to select the appropriate informants, as not all members of a society will be experts in all areas of local ecological knowledge, despite the awesome capacities of some individuals, and assembled bodies of knowledge generally require the synthesis of information from many local expert sources.

Where possible, informants should be screened for reliability and their information cross-checked by trained social scientists.

Moreover, people's willingness to discuss and share their knowledge varies immensely within and among societies, depending on the type of knowledge in question. This reflects an understanding of knowledge as power or privilege, especially in socially stratified societies. Further, when market forces intrude, or where individual accumulation is not sanctioned, knowledge is the key to an individual's or community's competitive performance, and thus it is unlikely to be shared widely.

Educational institutions should offer courses on local knowledge, because it is being rapidly eroded by Westernization, urbanization, monetization, and the alienation of youngsters from their traditions as a result of Western-style education. And those youngsters will eventually have to manage their local resources. They will find their formal education of little use for this, unless it is complemented by a firm grasp of local knowledge based on the experience of their elders. Thus, courses that blend both local knowledge and modern concepts of resources management should be made mandatory for students in educational institutions, particularly in the Third World.

In an effort to counteract the tacit negation of local knowledge by neglect, there has been an attempt on Tokelau (among other places in the Pacific) to document local ecological knowledge of fisheries and to introduce courses on it, taught by the elders, in primary and secondary schools (Toloa, Gillett, and Pelasio 1991). However, this was opposed by the Westernized elite, who perceived the Western curriculum to be superior and who would have lost status by having formally uneducated fishers teach classes, and thus the effort to see the fishers' local knowledge taught in the schools eventually failed. The elite, many of whom were educated in New Zealand, have a greatly diminished respect for the traditional system, the wrath of which they can escape by leaving Tokelau again.

To counteract these tendencies, some islanders believe in a need to restore the authority of the Council of Elders, either by restricting it to its traditional roles or by having a Tokelauan with a background in fisheries biology act as an advisor to the council, as well as by establishing more effective punitive measures for both traditional and modern offenses. There is a recognized need to more fully integrate biological information from stock assessment studies with local knowledge and to modify the age at which traditional education begins, so as to ensure the transmission of knowledge prior to the departure of youngsters for formal education overseas. There is also a need to convince the Westernized elite of the value of local knowledge. The value of this knowledge is undisputed by the residents, as is the need to adapt the traditional framework to the contemporary world and to establish mechanisms for blending biological information with local knowledge (Toloa, Gillett, and Pelasio 1991).

Tokelau provides but one small example of a phenomenon occurring throughout the tropics. Given the resurgence of ethnic pride in many areas, as among the aborigines of Australia, the Kanaks of New Caledonia, the Maoris of New Zealand, the First Nation peoples of Canada, and many others, it is hoped that this process will accelerate and become the norm.

Lessons for Modern Fisheries Management

1. Because sets of rights and rules that constitute folk-managed fisheries derive directly from local knowledge and concepts of resources on which the fishery is based, bodies of local knowledge are themselves a potentially valuable resource for use in modern fisheries management, and especially in co-management.

2. Practical, fish-behavior-oriented local knowledge, which focuses on the economically most important species, can provide a particularly important information base for managing tropical multispecies and multigear fisheries and their habitats,

because scientific knowledge of tropical inshore fisheries is relatively poor, and data required for conventional management are usually either scanty or nonexistent.

3. Local knowledge can provide a shortcut to pinpoint essential applied research needs, especially in localities where a traditional conservation ethic exists.

4. Because men and women often possess exclusive and perhaps complementary sets of local knowledge, an understanding of both is essential if local knowledge is to be useful in comprehensive fisheries management.

5. In documenting and analyzing local knowledge for application in modern fisheries management, it is imperative to go beyond a mere description of local nomenclature and classification of physical and biological environments and of marine fauna and flora. This is because the myriad details of local knowledge and their application have emerged from and function only within specific local contexts.

Notes

1. I make no apologies for drawing closely on Berger and Luckmann (1984) in this section, as elements of their important work provide a sorely needed conceptual framework for understanding the fundamental sociocultural importance of local knowledge. Such an understanding is of critical import in the practical management of traditional fisheries.

2. This illustrates a typical problem in attempting to assess the scientific validity of local knowledge, because although intuitively appealing, this perception is incorrect. Decreasing mean size does not necessarily indicate overfishing, as mean size can be expected to fall when effort is increasing. When a fish population has become exploited to a biological maximum sustainable level, mean sizes will be significantly smaller than when exploitation was at a low level. Whereas fishers may perceive this to indicate overfishing, in reality the fishery is approaching its most productive state. It is only when mean size falls below a certain limit that overfishing occurs. It is impossible without expensive and detailed scientific studies to determine where this point is. Thus, biologists claim correctly that such studies are needed because fishers' intuitive feelings can be very misleading. The problem is that the necessary

research is so time-consuming and expensive that the costs dwarf the benefits, except in large fisheries (R. E. Johannes, personal communication).

3. This is not intended to distinguish between conservation based on a desire to protect resources owing to their practical value from that owing to their perceived intrinsic value.

References

Akimichi, T.
   1978   The Ecological Aspect of Lau (Solomon Islands) Ethnoichthyology. *Journal of the Polynesian Society* 87:301–326.

Akimichi, T., and S. Sauchomal
   1982   Satawalese Fish Names. *Micronesica* 18:1–33.

Baines, G.B.K.
   1985   A Traditional Base for Inshore Fisheries Development in the Solomon Islands. In: K. Ruddle and R. E. Johannes (eds.), *The Traditional Knowledge and Management of Coastal Systems in Asia and the Pacific.* Jakarta, Indonesia: UNESCO. Pp. 39–52.

Berger, P., and T. Luckmann
   1984   *The Social Construction of Reality: A Treatise in the Sociology of Knowledge.* Harmondsworth, England: Penguin Books.

Berkes, F.
   1987   Common-Property Resource Management and Cree Indian Fisheries in Subarctic Canada. In: B. J. McCay and J. M. Acheson (eds.), *The Question of the Commons: The Culture and Ecology of Communal Resources.* Tuscon: University of Arizona Press. Pp. 66–91.

Borofsky, R.
   1987   *Making History: Pukapukan and Anthropological Constructions of Knowledge.* Cambridge, England: Cambridge University Press.

Brightman, R. A.
   1987   Conservation and Resource Depletion: The Case of the Boreal Forest Algonquians. In: B. J. McCay and J. M. Acheson (eds.), *The Question of the Commons: The Culture and Ecology of Communal Resources.* Tuscon: University of Arizona Press. Pp. 121–141.

Carrier, J. G.
   1987   Marine Tenure and Conservation in Papua New Guinea: Problems in Interpretation. In: B. J. McCay and J. M. Acheson (eds.), *The Question of the Commons: The Culture and Ecology of Communal Resources.* Tuscon: University of Arizona Press. Pp. 142–167.

Chapman, M. D.
    1987   Traditional Political Structure and Conservation in Oceania. *Ambio*
           16(4):201–205.

Cordell, J. C.
    1974   The Lunar-Tide Fishing Cycle in Northeastern Brazil. *Ethnology* 12:379–
           392.
    1978   Carrying Capacity Analysis of Fixed-Territorial Fishing. *Ethnology* 17:1–
           24.
    1990   Ed. *A Sea of Small Boats.* Cultural Survival Report 26. Cambridge, Mass.:
           Cultural Survival.

Dahl, A. L.
    1989   Traditional Environmental Knowledge and Resource Management in
           New Caledonia. In: R. E. Johannes (ed.), *Traditional Ecological Knowl-*
           edge: A Collection of Essays. Gland, Switzerland: International Union for
           the Conservation of Nature. Pp. 45–53.

Evans, D. W., and P. V. Nichols
    n.d.   *The Baitfishery of Solomon Islands.* Honiara, Solomon Islands: Ministry
           of Natural Resources.

Firth, R.
    1936   *We the Tikopia.* London: Allen and Unwin.

Forman, S.
    1967   Cognition and the Catch: The Location of Fishing Spots in a Brazilian
           Coastal Village. *Ethnology* 6:417–426.

Frusher, S. D., and S. Subam
    1984   Traditional Fishing Methods and Practises in the Northern Gulf of
           Papua. In: N. J. Quinn, B. Kojis, and P. R. Warphela (eds.), *Subsistence*
           Fishing Practises of Papua New Guinea. Lae, Papua New Guinea: Appro-
           priate Technology Development Institute. Pp. 80–88.

Gladwin, H.
    1980   Indigenous Knowledge of Fish Processing and Marketing Utilized by
           Women Traders of Cape Coast, Ghana. In: D. Brokensha (ed.), *Indige-*
           nous Knowledge Systems and Development. Washington, D.C.: University
           Press of America. Pp. 131–150.

Goldman, I.
    1970   *Ancient Polynesian Society.* Chicago: University of Chicago Press.

Hardin, G.
    1968   The Tragedy of the Commons. *Science* 162:1234–1248.

Hooper, A.
    1985   Tokelau Fishing in Traditional and Modern Contexts. In: K. Ruddle and
           R. E. Johannes (eds.), *The Traditional Knowledge and Management of*
           Coastal Systems in Asia and the Pacific. Jakarta, Indonesia: UNESCO. Pp.
           213–240.

Hora, S. L.
  1935 Ancient Hindu Conception of Correlation Between Form and Locomotion of Fishes. *Journal of the Royal Asiatic Society of Bengal: Science* 1:1–7.
  1948 Knowledge of the Ancient Hindus Concerning Fish and Fisheries of India: 1. References to Fish in Arthasastra (ca. 300 B.C.). *Journal of the Royal Asiatic Society of Bengal: Science* 14:7–10.
  1952 Fish in the Ramayana. *Journal of the Royal Asiatic Society of Bengal: Letters* 18(2):63–69.
  1953 Knowledge of the Ancient Hindus Concerning Fish and Fisheries of India: 4. Fish in the Sutra and Smrti Literature. *Journal of the Royal Asiatic Society of Bengal: Letters* 19(2):63–77.

Howard, A.
  1972 Polynesian Social Stratification Revisited: Reflections on Castles Built of Sand (and a Few Bits of Coral). *American Anthropologist* 74:811–823.
  1973 Education in 'Aina Pumehana': The Hawaiian-American Student as Hero. In: S. Kimball and J. Burnett (eds.), *Learning as Culture.* Seattle: University of Washington Press. Pp. 115–130.

Hviding, E.
  1988 *Marine Tenure and Resource Development in Marovo Lagoon, Solomon Islands: Traditional Knowledge, Use, and Management of Marine Resources, With Implications for Contemporary Development.* FFA Report No. 88/35. Honiara, Solomon Islands: South Pacific Forum Fisheries Agency.

Johannes, R. E.
  1978 Traditional Marine Conservation Methods in Oceania, and Their Demise. *Annual Review of Ecology and Systematics* 9:349–364.
  1981 *Words of the Lagoon: Fishing and Marine Lore in the Palau District of Micronesia.* Berkeley: University of California Press.
  1982 Traditional Conservation Methods and Protected Marine Areas in Oceania. *Ambio* 11(5):258–261.
  1989 Introduction. In: R. E. Johannes (ed.), *Traditional Ecological Knowledge: A Collection of Essays.* Gland, Switzerland: International Union for the Conservation of Nature. Pp. 5–8.

Johannes, R. E., and E. Hviding
  1987 *Traditional Knowledge of Marine Resources of the People of Marovo Lagoon, Solomon Islands, With Comments on Marine Conservation.* Technical Report to the Commonwealth Science Council, London.

Johannes, R. E., and W. MacFarlane
  1991 *Traditional Fishing in the Torres Strait Islands.* Canberra: Australian Department of Primary Industries and Energy.

Johannes, R. E., and K. Ruddle
  n.d. Human Interactions in Tropical Coastal and Marine Areas: Lessons From Traditional Resource Use. *Shoreline and Coastal Management.*

Kainang, A. J.
   1984 Traditional Fishing Technology of Yuo Island, East Sepik Province. In: N. J. Quinn, B. Kojis and P. R. Warphela (eds.), *Subsistence Fishing Practises of Papua New Guinea.* Lae, Papua New Guinea: Appropriate Technology Development Institute. Pp. 42–50.

Lambert, Le Père
   1900 *Moeurs et superstitions des Neo-Caledonniens.* Publication No. 14. Société d'Études Historiques. New Caledonia: Noumea. (Reprinted 1980.)

Levy, R.
   1973 *Tahitians: Mind and Experience in the Society Islands.* Chicago: University of Chicago Press.

Lingenfelter, S. G.
   1975 *Yap: Political Leadership and Culture Change in an Island Society.* Honolulu: University Press of Hawaii.

Malinowski, B.
   1935 *Coral Gardens and Their Magic.* London: Allen and Unwin.

Marcus, G.
   1978 Status Rivalry in a Polynesian Steady-State Society. *Ethos* 6:242–269.

McCoy, M. A.
   1974 Man and Turtle in the Central Carolines. *Micronesica* 10(2):207–221.

McGoodwin, J. R.
   1984 Some Examples of Self-Regulatory Mechanisms in Unmanaged Fisheries. FAO Fisheries Report 289, Supplement 2:41–61. Rome: FAO.

Morrill, W. T.
   1967 Ethnoichthyology of the Cha-Cha. *Ethnology* 6:405–416.

Nietschmann, B.
   1985 Torres Strait Islander Sea Resource Management and Sea Rights. In: K. Ruddle and R. E. Johannes (eds.), *The Traditional Knowledge and Management of Coastal Systems in Asia and the Pacific.* Jakarta, Indonesia: UNESCO. Pp. 125–154.
   1989 Traditional Sea Territories, Resources, and Rights in Torres Strait. In: J. C. Cordell (ed.), *A Sea of Small Boats.* Cambridge, Mass.: Cultural Survival. Pp. 60–93.

Norem, R. H., R. Yoder, and Y. Martin
   1989 Indigenous Agricultural Knowledge and Gender Issues in Third World Agricultural Development. In: D. M. Warren, L. J. Slikkerveen, and S. O. Titilola (eds.), *Indigenous Knowledge Systems: Implications for Agriculture and International Development.* Studies in Technology and Social Change No. 11. Ames: Technology and Social Change Program, Iowa State University. Pp. 91–100.

Polunin, N.V.C.
    1984 Do Traditional Marine "Reserves" Conserve? A View of Indonesian and
          Papua New Guinean Evidence. In: K. Ruddle and T. Akimichi (eds.),
          Maritime Institutions in the Western Pacific. Senri Ethnological Studies 17.
          Osaka, Japan: National Museum of Ethnology. Pp. 267–283.

Quinn, N. J., B. Kojis, and P. R. Warphela (eds.)
    1984 *Subsistence Fishing Practises of Papua New Guinea.* Lae, Papua New
          Guinea: Appropriate Technology Development Institute.

Raychaudhuri, B.
    1980 *The Moon and the Net: Study of a Transient Community of Fishermen at
          Jambudwip.* Calcutta: Anthropological Survey of India.

Ritchie, J., and J. Ritchie
    1979 *Growing Up in Polynesia.* Sydney: Allen and Unwin.

Ruddle, K.
    1987 *Administration and Conflict Management in Japanese Coastal Fisheries.*
          Fisheries Technical Paper 273. Rome: FAO.
    1991 The Transmission of Traditional Ecological Knowledge. Paper presented
          at the Second International Conference of the Association for the Study
          of Common Property, University of Manitoba, Winnipeg, September.

Ruddle, K., and T. Akimichi
    1984 Introduction. In: K. Ruddle and T. Akimichi (eds.), *Maritime Institutions
          in the Western Pacific.* Senri Ethnological Studies 17. Osaka, Japan:
          National Museum of Ethnology. Pp. 1–9.
    1989 Sea Tenure in Japan and the Southwestern Ryukyus. In: J. C. Cordell
          (ed.), *A Sea of Small Boats.* Cambridge, Mass.: Cultural Survival.

Ruddle, K., and R. A. Chesterfield
    1977 *Education for Traditional Food Procurement in the Orinoco Delta.* Ibero-
          Americana 53. Berkeley: University of California Press.

Shore, B.
    1982 *Sala'ilua: A Samoan Mystery.* New York: Columbia University Press.

Slikkerveen, L. J.
    1989 Changing Values and Attitudes of Social and Natural Scientists Towards
          Indigenous Peoples and Their Knowledge Systems. In: D. M. Warren, L.
          J. Slikkerveen, and S. O. Titilola (eds.), *Indigenous Knowledge Systems:
          Implications for Agriculture and International Development.* Studies in
          Technology and Social Change No. 11. Ames: Technology and Social
          Change Program, Iowa State University. Pp. 121–137.

Thrupp, L. A.
    1988 *The Political Ecology of Pesticide Use in Developing Countries: Dilemmas in
          the Banana Sector of Costa Rica.* Brighton, England, University of Sussex,
          Institute of Development Studies, Ph.D. diss.

Toloa, F., R. Gillett, and M. Pelasio
    1991   Traditional Marine Conservation in Tokelau. Paper presented to the
           South Pacific Commission, Twenty-Third Regional Technical Meeting
           on Fisheries, Noumea, New Caledonia, August 5–9, 1991.

Townsend, R., and J. A. Wilson
    1987   An Economic View of the Tragedy of the Commons. In: B. J. McCay and
           J. M. Acheson (eds.), *The Question of the Commons: The Culture and
           Ecology of Communal Resources.* Tuscon: University of Arizona Press. Pp.
           327–343.

Warren, C.A.B.
    1988   *Gender Issues in Field Research.* Newbury Park, Calif.: Sage Publications.

Warren, D. M.
    1989   The Impact of Nineteenth Century Social Science in Establishing Nega-
           tive Values and Attitudes Towards Indigenous Knowledge Systems. In:
           D. M. Warren, L. J. Slikkerveen, and S. O. Titilola (eds.), *Indigenous
           Knowledge Systems: Implications for Agriculture and International Devel-
           opment.* Studies in Technology and Social Change No. 11. Ames: Tech-
           nology and Social Change Program, Iowa State University. Pp. 171–183.
    1991   *Using Indigenous Knowledge in Agricultural Development.* World Bank
           Discussion Papers 127. Washington, D.C.: The World Bank.

Yamelu, T.
    1984   Traditional Fishing Technology of Bwaiyowa, Fergusson Island, Milne
           Bay Province. In: N. J. Quinn, B. Kojis, and P. R. Warphela (eds.),
           Subsistence Fishing Practises of Papua New Guinea. Lae, Papua New
           Guinea: Appropriate Technology Development Institute. Pp. 52–63.

8

Environmental Disaster and Fishery Co-Management in a Natural Resource Community: Impacts of the *Exxon Valdez* Oil Spill

Duane A. Gill

Introduction

On March 24, 1989, the supertanker *Exxon Valdez* ran aground in Prince William Sound, Alaska, spilling approximately eleven million gallons of North Slope crude oil. A massive oil slick formed and washed ashore on the western edge of the sound days later. Oil made its way south into the Gulf of Alaska, eventually oiling more than 1,300 miles of Alaskan coastline. This environmental disaster had a devastating effect on the natural environment and the human communities that rely on the area's renewable natural resources.

Major funding for this research was provided by grants from the National Science Foundation (DPP-9101093) and the Natural Hazards Research and Applications Information Center, University of Colorado (Boulder). This research was also supported by resources provided by the Coastal Research and Development Institute, University of South Alabama; The Social Science Research Center and Mississippi Agricultural and Forestry Experiment Station (Project No. MIS-4315), Mississippi State University; and the Fisheries Art Collective, Santa Cruz, California. Special recognition is due to fellow scientists on the research field team: J. Steven Picou and Christopher L. Dyer. Acknowledgments are also given to Arthur Cosby, Wolfgang Frese, Jon Carr, Maurie Cohen, and Pat Picou for their respective contributions to the completion of the study. On-site field support was provided by the Prince William Sound Science Center and the Copper River Delta Institute. The author remains solely responsible for the content of this chapter.

In particular, the region's commercial fisheries were severely impacted.

Alaskan commercial fishing is based on co-management. Fishers' organizations work with state and federal agencies in formulating and implementing fishery policies. Current policies divide the state into various regulatory areas. A diverse number of commercial fisheries exist in each regulatory area, based on locality (riverine or marine area), target species, and gear type (e.g., gill net, seine). Permits are required to fish in a particular fishery within a regulatory area, and a limited number of permits are issued for each fishery. Individual fisheries are further regulated by restrictions on when and where fish harvesting can take place. The regulatory area directly affected by the *Exxon Valdez* oil spill was Area E, which covers the Prince William Sound and Copper River Delta bioregion.[1]

Fisheries co-management is rooted in community values, orientations, and perceptions. In Prince William Sound, co-management originated from local folk management traditions of Native subsistence culture. Folk management was based on the communities' inextricable link to the environment for survival and existence. Communities in the bioregion are still inextricably linked to natural resources and are active in co-managing resources. The cultural orientation of contemporary co-management communities is influenced by Western culture tempered by a blend of Native traditions and a bond to the environment.

This chapter assesses impacts of an environmental disaster, the *Exxon Valdez* oil spill, on fisheries co-management in the community of Cordova, Alaska.[2] Historically, Cordova has been a center for fishery co-management activities in the bioregion as well as the state. The chapter will explore some of the linkages between the community and fisheries co-management, with a particular focus on community impacts of the disaster.

Impacts of an environmental disaster are best understood in the context of the bioregional ecology and cultural history of the community. Thus, the chapter begins with a brief description of the natural and human ecology of the bioregion. This is followed by a sociocultural profile of Cordova and an overview of fishery co-management activities. Next, a brief description of the oil spill and cleanup is presented, along with a description of some of the environmental consequences. Community impacts associated with the disaster are discussed next. Finally, some implications for co-management communities are identified and discussed.

Ecology of the Bioregion

Natural Ecology

Prince William Sound and the Copper River Delta constitute one of the world's most spectacular bioregions.[3] Rugged mountains capped with permanent ice fields provide the bioregion with natural boundaries to the north, east, and west, while the Gulf of Alaska borders the south. Large populations of several animal species annually migrate to the bioregion, which is rich in estuarine and marine food supplies. Annual salmon runs and the abundance of marine mammals gave rise to a 7,000-year-old subsistence culture among the Chugach Eskimo and Eyak natives.

The bioregion is located between $60°$ and $62°$ north latitude and $144°$ and $147°$ longitude. The area experiences nearly twenty-four hours of continuous daylight during the summer solstice and nearly twenty-four hours of continuous darkness during the winter solstice. The bioregion has a maritime climate characterized by heavy precipitation and moderate temperatures throughout the year. Vegetation generally consists of western hemlock and Sitka spruce, alpine tundra, and wet tundra. A variety of fish and shellfish are found in the bioregion, including salmon (king, silver, pink, red, and chum), herring, cod, halibut, crab (tanner, king, dungeness), shrimp, clams, and mussels. The bioregion is a habitat for marine mammals such as whales (finback, humpback, and gray), sea lions, harbor seals, and sea otters. The bioregion is inhabited by a variety of birds, including bald eagles, ravens, and jays. The coastal and inland habitat supports bears (brown and black), moose, deer, wolves, mountain goats, beavers, mink, martens, and weasels.

Prince William Sound is encompassed by the northern reaches of North America's temperate rain forest. It is characterized as a fjord-type estuary consisting of more than 2,000 miles of intricate shoreline formed by bays, fjords, islands, and tidewater glaciers. The sound's coastal forest strip maintains high water quality in areas where fish spawn and are reared. The cold waters of the sound are among the world's most biologically productive. When exposed to prolonged daylight in the spring and summer, highly oxygenated, high-nutrient waters sustain a complex ecological network.

The Copper River Delta encompasses 700,000 acres of pristine wetlands and is the largest contiguous wetland on the Pacific coast. It is composed of estuaries, muskeg, mudflats, and channels, which provide a rich habitat for a diversity of species. Nestled between the rugged Chugach and Wrangell–St. Elias mountain ranges, the delta is fed by the Copper River and several streams and other rivers that constitute a total watershed of 24,000 square miles. The delta fans out in a 65-mile arc into the Gulf of Alaska, where it is broken up by a series of barrier islands and outer sandbars. Commercial species of salmon and herring are found in the delta. Red salmon from the Copper River typically set the world market price for this commercial species.

Salmon are an integral part of the bioregion's ecological cycle. The life cycle of salmon is a focal point of the natural and cultural cycles of the bioregion. Salmon lay their eggs in gravel nests throughout the bioregion's streams and lakes. Salmon fry spend time in fresh water prior to heading to the open ocean to feed until they reach maturity. Upon full maturity, salmon return to their natal streams to spawn and die, thus renewing the cycle. The amount of time spent in fresh water and in the ocean to maturity varies depending upon the particular subspecies. During their life cycle, salmon provide sustenance to a wide variety of life, including other fish, birds, bears, and humans. Salmon have social and economic importance in every community in the bioregion.

Human Ecology

Prince William Sound is host to five human communities: Chenega Bay, Tatitlek, Whittier, Valdez, and Cordova. To varying degrees, these communities are reliant on the bioregion's renewable natural resources and can be classified as a natural resource community (NRC). An NRC is "a population of individuals living within a bounded area whose primary cultural existence is based on the utilization of renewable natural resources."[4] The predominate cultural relationship with the environment has evolved from the traditional subsistence culture of Alaskan Natives.

The subsistence culture of Alaskan Natives is based on the abundant renewable resources of the bioregion. Cultural cycles of anticipation, preparation, harvest, and utilization coincide with natural seasons and cycles of resource availability. Social and economic relationships in Native communities are shaped by folk traditions associated with natural resource utilization. A subsistence lifestyle creates patterns of family activities, religious and social ceremonies, cooperative social networks, and cultural identity based on the harvesting and sharing of natural resources.[5]

Cordova: A Profile of a Natural Resource Community

Cordova is a small, rural fishing community isolated from other communities by mountains, glaciers, rivers, and the sea. The community is located in an ecozone between Prince William Sound and the Copper River Delta. Historically, Cordova's beginnings are traced to four Eyak Native villages. Eyak villages in the region were exploited by Russian fur traders, and consequently, their populations were decimated. (As of 1990, there were only two living speakers of the Eyak language.) During the 1880s, salmon fishing and processing grew, attracting a stable population to the area. Four canneries were established in the vicinity of present-day Cordova.

In 1900, significant deposits of copper ore were discovered in the Wrangell Mountains to the north. The name "Cordova" was given to the site designated to be the point of transfer for the ocean transportation of ore mined in the interior. Cordova was linked by railroad to the mines, and copper mining flourished in the area for three decades. Although local fishing enterprises continued during this time, growth in commercial fishing coincided with the demise of copper mining in 1939.

After the closing of the mine, Cordova's population stabilized at about 1,000. During the 1970s Cordova's population doubled due to diversification in the commercial fishing industry and a generally high immigration rate for the state. Cordova now has a base population of 2,500, which nearly doubles during the summer fishing season. The majority of the residents are Anglo, as are most of the fishers. Alaskan Natives (comprising Eyak, Aleut, and other groups) constitute about 20 percent of the population and are highly integrated into the community and fishing industry. There is also a smaller, less integrated Oriental population, many of whom are employed in local canneries.

Commercial fishing dominates the economy of Cordova. Cordovan fishers own 55 percent of all Area E salmon fishery permits and 44 percent of all Area E herring fishery permits. Other commercial fisheries include cod, halibut, rockfish, sablefish, crab, and shellfish. More recently, commercial interests have developed in a pound-net fishery and an oyster fishery.

In addition to hosting the region's largest commercial fishing fleet, Cordova is home to several canneries, various support businesses (e.g., electronics, refrigeration, machinist, mechanic, net mending), and fishing organizations. The fishing industry provides the basis for locating various government agency operations in Cordova (e.g., U.S. Forest Service, U.S. Coast Guard, Alaska Department of Fish and Game). As Payne observed, "Nowadays, many would argue that without the fishing industry Cordova could not persist. Employment and other economic activities are so bound up in fishing that the very essence of the town depends on the fishery."[6]

Cordova maintains a subsistence heritage that is interwoven with the commercial fishing economy. The majority of Cordova's residents engage in subsistence activities that include harvesting and sharing fish, shellfish, moose, deer, berries, and so on. In a recent sample of Cordova households, 90 percent reported receiving subsistence resources, and households averaged more than 400 pounds of harvested resources.[7] As is found in most subsistence cultures, resource sharing in Cordova provides a basis for establishing and maintaining social relationships. However, subsistence is similar to the cash economy in that both are highly dependent on fishing.

Like other communities dependent on renewable natural resources, Cordova has developed a cultural cycle that corresponds with the biological cycle of fish, particularly salmon (see Figure 8.1). Although fishing is a year-round activity, the primary fishing season coincides with the season of highest biological productivity: March through September. In early February, the community symbolically awakens from the long winter with a celebration: the Cordova Iceworm Festival. Ceremonies include a blessing of the fleet, a parade, a local talent show, and a food fair. The festival marks the beginning of the preparation stage. Preparation involves readying the boat and gear, selecting and hiring crews, and identifying potential fishing areas for the upcoming season. This stage is characterized by a high level of optimism regarding the success of the impending harvest season.

The harvest cycle begins with the first herring runs in March. Harvesting continues with red salmon runs (May through June), pink salmon runs (July through August), and silver salmon runs (August through September). The harvest season continues for many year-round residents with the subsistence hunting of deer, moose, and waterfowl.[8] Utilization involves activities that convert the harvest into usable products, as well as the consumption of

Anticipatory Utilization Cycle

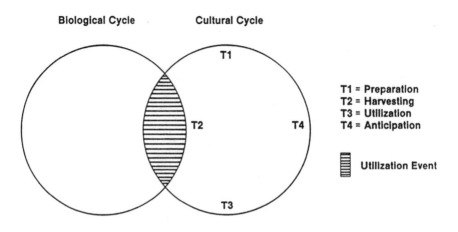

Figure 8.1 The cyclic relationship between cultural and natural events in a natural resource community.

these products. Utilization coincides with harvest (e.g., canning, smoking, and preserving fish and game as they are harvested) and extends into the fall and winter. The winter is marked by a period of reflection and anticipation. During this resting period, the previous harvest season is assessed, and plans for the upcoming season are contemplated and initiated.

The intertwinement of natural and cultural cycles leaves an imprint on the local inhabitants. They are acutely aware of their environment and the seasonal cycle. They become familiar with the currents of Prince William Sound and the shifting channels in the Copper River Delta. They develop a respect for the voracity of nature and an appreciation for the bounty nature provides. More importantly, they recognize the importance of stewardship. Stewardship forms a basis for co-management activities.

Fisheries Co-Management in a Natural Resource Community

As an environmentally aware community dependent on the natural resources of the bioregion, Cordova has traditionally been a leader in fisheries co-management. Commercial salmon fishing in the bioregion is 100 years old. In the beginning, commercial fishing had little impact on fishery populations, but as technology improved and fishers' numbers grew, signs of species stress and decline were observed. Fishers' organizations played an instrumental role in working with government agencies to co-manage the bioregion's fisheries.

Beginning with various fishers' unions in the 1920s and 1930s, fishers have been organized to have a voice in their business affairs. The Cordova District Fisheries Union (CDFU) emerged as fishers' representative in negotiations with canneries. The union was forced to disband in 1955 after a Department of Commerce ruling based on the Sherman Anti-Trust Act. However, the act did allow fishers to be organized as a marketing association through the Cordova Aquatic Marketing Association. CDFU continued as a representative in political and management issues and later changed its name to Cordova District Fishermen United.

CDFU has been an organizational leader in fishery co-management. When species stress and decline were observed, CDFU worked with branches of the state and federal government to manage the fishery. CDFU was a powerful entity during Alaska's transition into statehood and the ensuing development of fishery policies. CDFU represented fishers in developing co-management practices such as establishing limited entry, gear restrictions, and area closures.[9]

Statewide, Alaska's salmon harvest reached a low point in the late 1960s and early 1970s and threatened a collapse of the entire commercial fishery. In response, the 1971 state legislature created the Fisheries Rehabilitation, Enhancement, and Development (FRED) Division to ensure the perpetual use of Alaska's aquatic resources. Additional legislation in 1974 provided for the establishment of private nonprofit hatchery programs. Cordova fishers initiated the formation of the Prince William Sound Aquaculture Corporation (PWSAC) in December 1974 to establish a hatchery program to produce an adequate supply of common-property fish for all user groups. The PWSAC structure included representatives from commercial, subsistence, recreational, and personal use fisheries, as well as from fish processors, Native villages and corporations, and local communities.[10] The financial structure of PWSAC is based on salmon-enhancement tax revenues collected from all salmon fishers and on proceeds from the sale of fish returning to the hatchery.

PWSAC reflects the nature of the co-management orientation in the Area E fishery. It was initiated by fishers who are integrated with the resource and recognize the up-and-down cycles of fishing. The hatcheries were built by the fishers, who donated time and materials. For example, the Armin F. Koernig hatchery was built in 1975 through a complete volunteer effort of commercial fishers. Currently, there are five hatcheries in Prince William Sound, three operated by PWSAC, one by the Valdez Fisheries Development Association, and one by the state's FRED Division. The effort to develop and enhance the relationship with the environment seems to have paid off.

As plans for hatcheries were being developed, another plan, the construction of a pipeline from the rich oil fields of the North Slope to a terminal in Valdez, began to unfold.[11] CDFU actively opposed the location of a pipeline terminal in Valdez because of the high risks associated with terminal activities and with tanker transportation in the environmentally rich waters of the bioregion. An alternative overland trans-Canadian pipeline was proposed. The CDFU joined forces with Alaskan Native and environmental groups, and together they forced the issue to the Supreme Court, which upheld the CDFU coalition's position. In late 1973, however, a special right-of-way legislation was introduced in the U.S. Congress to circumvent the court rulings. In the U.S. Senate, the vote was a 49–49 tie, which was broken in favor of the Valdez terminal plan by Vice President Spiro T. Agnew. A similar bill passed the U.S. House, and President Nixon signed the Pipeline Authorization Act on November 16, 1973.

Payne's description of the organization dynamics of CDFU and the community of Cordova during the pipeline controversy reveals four fundamental orientations toward resource management.[12] First, fishers are keenly aware of the local ecology and potential threats to their natural resource base. The fishery permit structure promotes a sense of ownership and stewardship over the economic base. They are in constant contact with the ecosystem.

Second, CDFU emerged out of a strong sense of solidarity vis-à-vis resource stewardship. Solidarity was enhanced by the democratic orientation of the organization and the respected leaders, who come from the ranks of the fishers themselves. The strong organization effectively channels the fishers' concern for managing fishery resources. Further, CDFU has been influential with fishers' organizations throughout the state in shaping co-management strategies and policies.

The nature of fishing also affects management orientations. Fishing promotes innovation to improve equipment and techniques and to make do with the resources at hand. Although fishing is an individualistic competitive activity, it also promotes cooperation. Fishing is dangerous work involving machinery, a threatening environment (sudden violent storms, cold, ice, wind, etc.), and a high degree of unpredictability. Each year is marred by occupational fatalities. Each individual fisher realizes the high likelihood of needing help someday, and thus, fishers as a group come to rely on each other.

Finally, the social fabric of a small community facilitates cooperative face-to-face relationships. The physical isolation leads to innovation, volunteerism, civic participation, and interest in community affairs. This fabric is further strengthened by subsistence networks that bind people to each other and the environment.

The community empowerment and co-management orientations of CDFU, PWSAC, and the Cordova community generated success. By the mid-1980s, commercial fishing appeared to be secure. Hatchery fish were plentiful, fishers were active in the management of the fishery, and they were earning a good income. But this security would be threatened at the end of the decade with an unprecedented environmental disaster: the *Exxon Valdez* oil spill.

The *Exxon Valdez* Oil Spill and Cleanup: A Brief Overview

The Oil Spill

The *Exxon Valdez* oil spill was a severe blow to the local fishery and its co-management structure.[13] The spill occurred at the beginning of the most biologically active season of the year, thus maximizing the threat to the bioregion. The 1,300 miles of impacted coastline included parts of Prince William Sound, Cook Inlet, and the Alaskan Peninsula. Kodiak Island was surrounded by oil. Coastal and subtidal habitats in these regions were exposed to oil and further disrupted by cleanup operations.

The oil spill had an immediate impact on the bioregion's marine and terrestrial life. Tens of thousands of birds died from exposure to oil and more than 1,000 otters were confirmed to have died. Mussels and clams were contaminated, as were various other marine subsistence resources. The oil contamination entered the food chain as predators and scavengers consumed carcasses tainted with oil. An extensive effort to remove dead birds and otters limited the impacts to the food chain, yet ecological damages did occur.

The fishery experienced disruption in the aftermath of the oil spill. Initial efforts were made by commercial fishers to secure the hatcheries, although it became apparent that hatchery fry were likely to be impacted. The fry and the oil would travel the same current out of the sound and into the Gulf of Alaska. Many fishers feared the impact on fisheries would be felt for years to come.

The Alyeska Pipeline Service Company, which was responsible for containing and cleaning up oil spills, was unprepared to handle a spill of such magnitude.[14] Further, the response was hampered by poor coordination among the corporate organizations, state and federal agencies, and local communities.

Controversy and outrage were generated within impacted communities and among the public regarding the slow response, effectiveness of cleanup techniques, and compensation strategies employed by the responsible corporate and government entities.[15] There was public sentiment that the spiller should not have been in charge of the cleanup effort. Local residents were incensed that their local expertise was ignored. Some felt cleanup efforts were futile, the job being too great to be undertaken by human hands. There were also concerns about the impacts of cleanup, secondary spills, and the disposal of waste beaches.

The Cleanup

Exxon and its primary contractor, Veco, took command of the cleanup operations and initiated a series of actions to contain and clean up the oil. These included subcontracting fishing vessels and hiring thousands of people to participate in the cleanup. Fishing vessels were used to lay containment booms around oil slicks while skimmer vessels collected the oil. Fishing vessels were also used to transport materials, wastes, and personnel. Oil was removed from beach surfaces by hand and by using cold and hot water from high-pressure hoses. Bioremediation processes, which entail the application of bacteria and fertilizer to oiled beaches in order to accelerate the natural breakdown of oil, and chemical dispersant were used in some areas.

Cleanup activities continued throughout the summer, ending in mid-September. At that point, nature was commissioned to clean the shores with its savage winter storms. Limited cleanup efforts would be made in subsequent years, but the ecological recovery was left to nature. However, there was uncertainty about the time needed for recovery.

Local communities experienced a boomtown phenomenon as an army of cleanup workers and media reporters inundated the area. The population of Valdez temporarily increased from 3,500 to more than 10,000 in two months. Cordova's population increased from 2,500 to more than 6,000. Population booms impacted housing, business patterns, mental health services, government services, and the judicial system. Community officials reported dramatic increases in violent crimes and drug and alcohol abuse.

The oil spill and cleanup disrupted community, family, and occupational lifestyles. For many residents, the spill lingers through litigation, poor fishing, and resource recovery uncertainty. Compared to the number of studies on ecological impacts, however, few studies were conducted on the impacts of the *Exxon Valdez* oil spill on humans. The critical question remains, What happened to the communities?

Analyzing the *Exxon Valdez* Oil Spill

Shrivastava's stakeholder model of industrial crises is an appropriate model for analyzing the *Exxon Valdez* oil spill.[16] Primary stakeholders in industrial crises are corporations, government, and affected communities. The public, public interest groups, and the mass media are secondary stakeholders. Each stakeholder operates from a self-interest frame of reference that determines its response to a crisis.[17] Primary stakeholders compete with each other to have their respective perspectives accepted by the public as "truth," with the corporate view of reality prevailing in most cases. The tendency in industrial crises is for corporations and government to minimize the extent of damage, appeal to scientific knowledge, and exclude victims from participating in critical decisions.[18]

Affected communities operate from a frame of reference based on personal experience and knowledge and are typically powerless to form a proactive response. Further, there may be a lack of community consensus as to the nature and extent of impact. These communities tend to vent anger, frustration, and bitterness toward the corporation and government.

The affected community is ultimately responsible for self-protection. Because corporations and government tend to resist changes in their frame of reference, even when such changes are in their best interests, community-based organizations must ultimately force such changes. However, communities and community-based organizations may be weakened in the aftermath of an industrial crisis.

Community Impacts of the *Exxon Valdez* Oil Spill

Industrial crises can be thought of as technological disasters. Social science research indicates a pattern of community disruption following a technological disaster. Because co-management is linked to the community, community disruption can lead to co-management disruption. Following is a brief summary of the literature on technological disasters.

An environmental disaster such as the *Exxon Valdez* oil spill poses problems that are unique when compared to natural disasters such as earthquakes, hurricanes, floods, and volcanic eruptions. Like other technological disasters, oil spills involve a *loss* of control over an activity that was perceived to be controllable, as opposed to natural disasters, which involve a *lack* of control over processes that are uncontrollable. Loss of control includes mechanical and structural failures as well as breakdowns in social organization.

Most communities are not experienced in or prepared for a technological disaster; thus, responses, information flows, and expert intervention patterns prove inadequate. Further, victims of technological disasters exhibit more anger, rage, and hostility than do victims of natural disasters. Social-psychological impacts tend to be more widespread and persistent in technological disasters than in natural disasters. Indeed, stress associated with technological disasters may actually increase over time. The initial technological disaster can be exacerbated by a series of secondary disasters fueled by litigation, uncertainty, and continued disruptions.[19]

In a natural disaster, social support networks and a therapeutic community atmosphere typically emerge, whereas a technological disaster may be accompanied by an abrasive community atmosphere. Internal segmentation of community groups typically develops along lines reflecting relative threats to property, livelihood, family, and the individual. Furthermore, "social conflict characterizes the 'spacial interaction' of *technology* and community as opposed to social cooperation characteristic of the 'spacial interaction' of *ecology* and community."[20]

The literature on technological disasters indicates that negative community impacts are the norm. When examining the community of Cordova, various types of impacts can be attributed to the *Exxon Valdez* oil spill. Community impacts can be broadly categorized as follows: economic disruption, fishery disruption, recovery uncertainty, community divisiveness, litigation, and social-psychological stress. As in other technological disasters, there is overlap between these categories, as well as interactive and cumulative effects.

Economic Disruption

Like other communities impacted by the oil spill, Cordova experienced tremendous economic upheaval. First, there was an economic bonanza as Exxon spent more than $2 billion in the cleanup effort. A significant portion of the money passed through the Prince William Sound economy. Some fishers were able to lease their boats to Veco at more than $1,500 per day to assist in the cleanup effort. This "money spill" created a temporary bonanza whereby some individuals made much more money cleaning up oil than they could have made from fishing. However, not all fishers received leases, or wanted them.

The local economy experienced occupational disruptions. The hourly wages for shoreline cleanup exceeded $16 an hour. With the fishing season in jeopardy, those relying on fishing-related occupations faced uncertain options. Fishing crew members and cannery workers had to decide between their occupation and cleanup employment. The $16-an-hour rate was significantly higher than most local jobs paid. Thus, a temporary labor shortage in lower-paying service occupations was created as people quit jobs to work on the cleanup.

The local economy was impacted further as traditional business patterns were interrupted. Buying and service patterns changed as oil spill cleanup temporarily replaced fishing as the dominant economic activity. The oil spill and cleanup also took a toll on the budgets of the town government, Native American–run corporations, and other established organizations as they incurred debt struggling to respond to the spill and cleanup.

Economic upheavals resulted in downward economic trends during 1990, 1991, 1992, and 1993 due to disruptions in the fishing industry. The price of fish dropped dramatically in 1990 and generally stayed depressed from 1990–1993.[21] Many individuals who made "spill" money in 1989 overinvested in their fishing business (new boats, equipment, etc.) and were unable to make the payments in the lean years that followed. Foreclosures and bankruptcies have recently become a more common occurrence. Occupational disruption continues as the future of fishing remains uncertain.

Fishing Disruption

The spill severely disrupted the 1989 commercial fishing season. The herring season was essentially closed, as was more than one-third of Prince William Sound fishing grounds. Those who chose to fish received a lower price for their product. In addition, there was concern that oil-contaminated fish might enter the market and damage public demand for the product.

The fisheries continue to be disrupted. Although 1990 was a banner harvest, the price for fish declined. The 1991 and 1992 seasons were marked by poor returns of herring and salmon and by low prices. The seine fishery for pink salmon was limited in 1991 and closed in 1992 due to weak returns. The 1993 herring season essentially was canceled. Although there is considerable scientific and public debate regarding the extent to which these seasonal disruptions are attributable to the oil spill, there is a strong perception among many fishers that the spill is the root cause of the problem.

Recovery Uncertainty

The patterns of disruption in the commercial fishery reinforce an underlying uncertainty regarding resource recovery. There is uncertainty about the oil's long-term effects on the region's natural resource base. There is uncertainty about whether the fish will come back, whether there will be enough fish, and how long it will take for the numbers to return to normal levels.

Science has been unable to adequately respond to critical information about bioregion impacts and recovery.[22] Experts have varying and conflicting reports regarding the degree of impact and recovery. Because they are unable to provide a consensus, there is a sense that scientific experts represent the interests of those who write the check rather than the interests of objective science. This fuels the uncertainty about resource recovery.

There is also a sense of uncertainty about community recovery. When will the local economy recover? When will community schisms heal? When will the litigation be settled? When will the aftershocks stop? When will it all be relegated to history?

Community Divisiveness

Every community is characterized by group conflict and fault lines where conflicting groups mesh. Conflict is ordered by consensus, which serves to diminish conflict by emphasizing the common good. Cordova contained preexisting conflict groups based on demographics (race, sex, occupation, age), attitudes (political, religious, regarding development), and residential status (permanent or seasonal, inside or outside city limits).

The oil spill weakened the social fabric of Cordova by creating deep schisms along various community fault lines. First, the community became divided because some received or accepted "Exxon money" and others did not or could not. People were cast as being greedy or labelled as "Exxon whores" or "spillionaires" as a result of their participation in the cleanup effort. This strained and split family relationships and long-term friendships. The community was further divided by a series of lawsuits within the town government that placed an additional financial burden on the town.

Another source of community conflict derives from a decision by the local Chugach Alaska Native Corporation to engage in clear-cut logging of local corporation land. The logging decision was spurred in part by financial problems incurred during the oil spill and the ensuing economic difficulties in the community. Many residents view logging as a threat to the fishing industry and an affront to a healthy bioregion. Yet, some feel a need to diversify the local economy through this type of development.

Finally, a twenty-five-year-old conflict regarding the construction of a road to Cordova has reemerged. Half of the community is for such a road, and the other half is against it. Proponents cite the economic benefits of cheaper transportation costs and the economic diversity of recreation and tourism industries. Opponents see the road as a threat to their traditional way of life and lament the potential of a steady stream of tourists in Winnebagos interrupting their way of life. Others are suspicious that the road might speed up resource extraction and pose further threats to the bioregion.

Litigation

Litigation activities are another source of community impact. More than two thousand lawsuits have been filed in the case. Several Cordova businesses and residents are involved in litigation activities. Legal action by the federal and state government against Exxon were settled in 1992. The settlement included the establishment of a $900 million restoration fund and a Trustee Council to develop and implement restoration plans.[23] Although intended to restore the community, the fund and council have generated controversy and conflict, further dividing the community.

Various class-action and individual lawsuits remain unresolved. Native corporations, fishers and their crews, and oil spill cleanup workers are three major class action groups. Individual lawsuits have been filed by fishers, business owners, and other residents impacted by the spill. The ongoing litigation creates additional uncertainty regarding individual and community recovery.

The high stakes have heightened the traditional antagonistic posturing found in our legal system. Subpoenas and depositions can be intimidating and can be used as "weapons" to discourage plaintiffs from pursuing the suit. Further, a corporate cost-benefit approach leads to a strategy of delaying resolution of litigation over a long period of time.[24] Such a tactic prolongs the secondary disaster, the negative social impacts to the community.

Social-Psychological Stress

The *Exxon Valdez* oil spill took an immediate emotional and psychological toll on people and communities. Pain, anger, frustration, depression, betrayal, anxiety, despair, sadness, and grief were common emotions experienced by residents of the bioregion. Tremendous stress was created as livelihoods were threatened or destroyed. The social fabric was torn by broken friendships and business relationships. There was a sense that a way of life had been lost.

Impacted communities reported increased incidences of alcohol and drug abuse, domestic violence, mental health problems, and occupation-related problems. Individuals in impacted communities experienced clinically measurable levels of generalized anxiety disorder, post-traumatic stress disorder, and depression.[25] These conditions placed increased demands on mental health services.

In Cordova, social-psychological stress was exacerbated by turnover in the community mental health administration and staff during this period of increased demand. Staff turnover in other community support agencies (e.g., law enforcement, town government) added to the sense of personal disruption.

Long-term social-psychological stress has been documented in the community and has actually increased over time among certain groups.[26] The combination of litigation, recovery uncertainty, and disruptions in the economy, fishery, and community maintains a higher-than-expected level of stress in Cordova.

In many respects the community is still grieving the environmental destruction and community disruption caused by the spill. Many individuals, particularly fishers and those connected to the industry, find it difficult to gain a sense of closure before the fisheries are clearly recovered and the litigation is resolved. This sets a stage for continued social-psychological stress within the community.

Summary of Community Impacts

Like other technological disasters, the Exxon Valdez oil spill had negative impacts on the community. Cordova experienced initial disruptions in its economy, commercial fishery, and way of life. Continuing economic and commercial fishing disruptions, along with uncertainty regarding resource recovery, contribute to a secondary disaster experience. Litigation provides another secondary disaster experience, as the legal process creates additional frustration and hostility for community victims. The abrasive community atmosphere resulting from the various schisms adds to the secondary disaster.

These community impacts have an affect on the ability to co-manage resources. The ongoing social disruption undermines the community's co-management network. This makes co-management more difficult and places additional stress on organizational leadership. Crises become politicized, with various special interests competing to influence the outcome. As a result, community leadership has different requirements placed upon it. One of the requirements is to respond to vulnerabilities exposed by the crisis.

Conclusions

The *Exxon Valdez* oil spill provides an important case study of environmental disaster impacts on a co-management community. The event created an abrasive community atmosphere characteristic of technological disasters. It also undermined trust between the local community and government, industry, and science. It underscored the conflict between state "progress" and local control. More importantly, the ecosystem, which is the base for community existence, was severely disrupted. These factors are threats to co-management and represent obstacles to recovery.

More generally, the oil spill provides an awareness of the need to think bioregionally in the co-management of resources. It demonstrates the vulnerability of single-resource economies and the inability of the scientific community to reach conclusions and consensus. The spill also shows that community empowerment is essential to obtaining greater control over resource management decisions.

Cordova continues to experience the *Exxon Valdez* disaster. The economic and fishing disruption, uncertainty, community segmentation, and social-psychological stress are predictable patterns of community impacts. There have been few attempts at cultural/community restoration. However, Cordova is a spirited community that has withstood previous disasters and crises. Recovery and restoration are difficult processes, but the community will prevail.

Lessons for Modern Fisheries Management

1. Natural resource communities should develop local strategies to ensure the protection of renewable resources against technological disasters.
2. Co-management communities need to be empowered by having a role in decisions regarding hazards and risks to their bioregion.
3. In the event of an environmental crisis, and where there is an absence of credible science, folk management strategies should be given particular attention.
4. In regions that are at risk for loss of natural resources, proactive diversification of the economy should be undertaken by local communities to provide an economic buffer in the event of a disaster.

Notes

1. "A bioregion is a part of the earth's surface whose rough boundaries are determined by natural rather than human dictates, distinguishable from other areas by attributes of flora, fauna, water, climate, soils and landforms, and the human settlements and cultures those attributes have given rise to. . . . The general contours of the regions themselves are not hard to identify, and indeed will probably be felt, understood or sensed, in some way known to many of the inhabitants, and particularly those still rooted in the land, farmers and ranchers, hunters and fishers, foresters and botanists, and most especially, across the face of America, tribal Indians, those still in touch with a culture that for centuries knew the earth as sacred and its well-being as imperative" (Sale 1974:227).

2. This description is based on information collected over the course of the four years since the oil spill. The author was part of a social science research team

that conducted research in Cordova with regard to community impacts of the oil spill. The study employed a variety of data collection methods such as ethnographic interviews, standardized household surveys, participant observation, archival searches, and retrieval of secondary data. Although at this time the research project is not complete, to date, the project has yielded five publications and a dissertation (Cohen 1993a, 1993b; Dyer, Gill, and Picou 1992; Dyer 1993; Picou and Gill 1993; Picou et al. 1992).

3. For a more detailed description of the regional natural and cultural ecology, see Thomas et al. (1991) and the Alaska Geographic Society (1993).

4. See Dyer, Gill, and Picou (1992).

5. Walter Meganack, traditional village chief of Port Graham Native Village, described the native natural resource community and the impacts the *Exxon Valdez* oil spill had on the Native way of life in an editorial entitled "The Day the Water Died." The text has been reprinted in Levkovitz (1990:44–45).

6. See Payne (in press).

7. See Stratton (1989).

8. Most hunting is done under the guise of recreation/sport, but the utilization and distribution of these resources typically follow subsistence networks.

9. A more detailed description of the legislative action toward limited entry and gear restrictions can be found in Cooley (1963), Kruse (1988), Morehouse and Hession (1972), Morehouse and Rogers (1980), and Royce (1989).

10. In 1976, the state of Alaska adopted the user-group structure established by PWSAC.

11. Payne (in press) provides a detailed description of the role of CDFU in the pipeline controversy.

12. See Payne (in press).

13. The spill was actually predicted hours before it occurred in a teleconference linkup with Mayor John Devens and a town meeting in Valdez. Cordova resident Riki Ott stated, "When, not if, the big one does occur and much or all of the income from a fishing season is lost, compensation for processors, support industries, and local communities will be difficult, if not impossible, to obtain" (Keeble 1991:24–25).

14. Indeed, at the time of the accident, Alyeska's primary response vessel was in dry dock, its supply of boom material and dispersant was inadequate, and only a handful of its personnel who were skilled in operating the needed equipment were available.

15. For a more complete description of the accident and cleanup activities, see Davidson (1990), Keeble (1991), and Lethcoe and Nurnberger (1989).

16. Shrivastava (1987). "An industrial crisis is a complex system of interdependent events and involves multiple conflicting stakeholders" (1987:19).

17. "A frame of reference is the method people or organizations use to select and process information" (Shrivastava 1987:87).

18. See Gephart (1984) and Shrivastava (1987).

19. For additional information on technological and natural disasters, see Ahearn (1981); Baum, Fleming, and Singer (1983); Couch and Kroll-Smith (1985); Edelstein (1988); Freudenburg and Jones (1991); Gill (1986); Gill and Picou (1989, 1991); Kroll-Smith and Couch (1991); Picou (1984); and Picou and Rosebrook (1993). For an alternative view espousing the existence of no distinguishable difference between technological and natural disasters, see Quarantelli (1985, 1987).

20. See Picou and Gill (1993).

21. As Cohen (1993b) notes, there was an economic recession at the national level during this time, as well as a general downturn in fisheries economics for the state.

22. For example, compare the proceedings from the Exxon Valdez Oil Spill Symposium in Anchorage Alaska (February 1993) and the Third Symposium on Environmental Toxicology and Risk Assessment: Aquatic, Plant, and Terrestrial, in Atlanta, Georgia (April 1993).

23. The $900 million will be paid over a ten-year period as specified in the settlement with the United States and the State of Alaska. As of 1993, approximately one-third of the fund had been spent.

24. It should be noted that litigation connected with the 1978 *Amoco Cadiz* oil spill remains unresolved as of this writing.

25. See Donald et al. (1990) and Palinkas et al. (1993).

26. See Picou and Gill (1993).

References

Ahearn, Frederick L.
1981 "Man-Made and Natural Disasters: A Comparison of Psychological Impacts." Unpublished paper presented at the annual meeting of the American Public Health Association, Los Angeles, CA.

Alaska Geographic Society
1993 *Alaska Geographic* 20(1).

Baum, Andrew, Raymond Fleming, and J. E. Singer
1983 "Coping With Victimization by Technological Disaster." *Journal of Social Issues* 39(2):117–138.

Cohen, M. J.
1993a "Economic Aspects of Technological Accidents: An Evaluation of the *Exxon Valdez* Oil Spill On Southcentral Alaska." Unpublished Ph.D. dissertation, Department of Regional Science, University of Pennsylvania, Philadelphia, PA.
1993b "Economic Impact of an Environmental Accident: A Time-Series Analysis of the *Exxon Valdez* Oil Spill in Southcentral Alaska." *Sociological Spectrum* 13(1):35–63.

Cooley, R. A.
1963 Politics and Conservation: The Decline of the Alaska Salmon. New York: Harper and Row.

Couch, S. R., and J. S. Kroll-Smith
1985 "The Chronic Technical Disaster: Toward a Social Scientific Perspective." *Social Science Quarterly* 166(3):564–575.

Davidson, A.
1990 In the Wake of the *Exxon Valdez:* The Devastating Impact of the Alaskan Oil Spill. San Francisco, CA: Sierra Club.

Donald, R., R. Cook, R. F. Bixby, R. Benda, and A. Wolf
1990 The Stress Related Impact of the Valdez Oil Spill on the Residents of Cordova and Valdez, Alaska. Valdez, AK: Valdez Counseling Center.

Dyer, C. L.
1993 "Tradition Loss as Secondary Disaster: Long-Term Cultural Impacts of the *Exxon Valdez* Oil Spill." *Sociological Spectrum* 13(1):65–88.

Dyer, C. L., D. A. Gill, and J. S. Picou
1992 "Social Disruption and the *Valdez* Oil Spill: Alaskan Natives in a Natural Resource Community." *Sociological Spectrum* 12(2):105–126.

Edelstein, M. R.
1988 Contaminated Communities: The Social and Psychological Impacts of Residential Toxic Exposure. Boulder, CO: Westview Press.

Freudenburg, William R., and Timothy R. Jones
1991 "Attitudes and Stress in the Presence of Technological Risk: A Test of the Supreme Court Hypothesis." *Social Forces* 69(4):1143–1168.

Gephart, R. P.
1984 "Making Sense of Organizationally Based Environmental Disasters." *Journal of Management* 10(2):205–225.

Gill, Duane A.
1986 "A Disaster Impact Assessment Model: An Empirical Study of a Technological Disaster." Unpublished Ph.D. dissertation, Department of Sociology, Texas A&M University, College Station, TX.

Gill, Duane A., and J. Steven Picou
  1989 "Toxic Waste Disposal Sites as Technological Disasters." Pp. 81-97 in D. Peck (ed.), Psychosocial Effects of Hazardous Toxic Waste Disposal on Communities. Springfield, IL: Thomas.
  1991 "The Social Psychological Impacts of a Technological Accident: Collective Stress and Perceived Health Risks." *Journal of Hazardous Materials* 27(1):77–89.

Keeble, J.
  1991 Out of the Channel: The Exxon *Valdez* Oil Spill in Prince William Sound. New York: HarperCollins.

Kroll-Smith, J. Stephen, and Stephen R. Couch
  1991 "What Is a Disaster? An Ecological-Symbolic Approach to Solving the Definitional Debate." *International Journal of Mass Emergencies and Disasters* 9(3):355–366.

Kruse, G. H.
  1988 An Overview of Alaska's Fisheries: Catch and Economic Importance of the Resources, Participants in the Fisheries, Revenues Generated, and Expenditures on Management. Fishery Research Bulletin 88-01. Juneau: Alaska Department of Fish and Game, Division of Commercial Fisheries.

Lethcoe, N. R., and L. Nurnberger (eds.)
  1989 Prince William Sound Environmental Reader 1989— *T/V Exxon Valdez* Oil Spill. Valdez, Alaska: Prince William Sound Conservation Society.

Levkovitz, T.
  1990 The Day the Water Died: A Compilation of the November 1989 Citizens Commission Hearings on the *Exxon Valdez* Oil Spill. Anchorage: Alaska Natural Resource Center, National Wildlife Federation.

Morehouse, T. A., and J. Hession
  1972 "Politics and Management: The Problem of Limited Entry." Pp. 279–331 in A. R. Tussing, T. A. Morehouse, and J. D. Babb, Jr., (eds.), Alaska Fisheries Policy: Economics, Resources, and Management. Fairbanks: Institute of Social, Economic, and Government Research, University of Alaska.

Morehouse, T. A., and G. W. Rogers
  1980 Limited Entry in the Alaska and British Columbia Salmon Fisheries. Fairbanks: Institute of Social, Economic, and Government Reseach, University of Alaska.

Palinkas, L. A., M. A. Downs, J. S. Petterson, and J. Russell
  1993 "Social, Cultural, and Psychological Impacts of the *Exxon Valdez* Oil Spill." *Human Organization* 52(1):1–13.

Payne, James T.

(In press) "Our Way of Life Is Threatened and Nobody Seems to Give a Damn": The Cordova District Fisheries Union and the Trans-Alaska Pipeline. Anchorage: Alaska Pacific University Press.

Picou, J. S.

1984 Ecological, Physical, Economic, Sociological, and Psychological Assessment of the Illinois Central Gulf Train Derailment. Vol. 5: Sociological Assessment. Baton Rouge, LA: Gulf South Research Institute.

Picou, J. S., and D. A. Gill

1993 "Disaster and Long-Term Stress: Impacts of the *Exxon Valdez* Oil Spill on Commercial Fishermen." Paper presented at the annual meeting of the Southwestern Sociological Association, New Orleans, LA.

Picou, J. S., D. A. Gill, C. L. Dyer, and E. W. Curry

1992 Disruption and Stress in an Alaskan Fishing Community: Initial and Continuing Impacts of the *Exxon Valdez* Oil Spill. *Industrial Crises Quarterly* 6(3):235–257.

Picou, J. S., and D. D. Rosebrook

1993 "Technological Accident, Community Class-Action Litigation, and Scientific Damage Assessment: A Case Study of Court-Ordered Research. *Sociological Spectrum* 13(1):117–138.

Proceedings

1993 Proceedings of the Exxon Valdez Oil Spill Symposium, Anchorage, AK. February.

1993 Proceedings of the Third Symposium on Environmental Toxicology and Risk Assessment: Aquatic, Plant, and Terrestrial, Atlanta, GA. April.

Quarantelli, E. L.

1985 "What Is a Disaster? The Need for Clarification in Definition and Conceptualization in Research." In Barbara J. Sowder (ed.), Disasters and Mental Health: Selected Comtemporary Perspectives. Pp. 41–73. U.S. Department of Health and Human Services, National Institute of Mental Health. Washington, DC: G.P.O.

1987 "What Should We Study? Questions and Suggestions for Researchers About the Concept of Disasters." *International Journal of Mass Emergencies and Disaster* 5(1):7–32.

Sale, K.

1974 The Schumacher Lectures. Vol. 2. London: Random Century.

Shrivastava, P.

1987 Bhopal: Anatomy of a Crisis. Cambridge, MA: Ballinger.

Stratton, L.

1989 Resource Use in Cordova: A Coastal Community of Southcentral Alaska. Technical Paper No. 153. Anchorage: Alaska Department of Fish and Game Division of Subsistence.

Thomas, G. L., E. H. Backus, H. H. Christensen, and J. Weigand
  1991   Prince William Sound/Copper River/North Gulf of Alaska Ecosystem. Cordova, AK: Prince William Sound Science Center, Copper River Delta Institute, and Conservation International.

9

Are Folk Management Practices Models for Formal Regulations? Evidence From the Lobster Fisheries of Newfoundland and Maine

Craig T. Palmer

Introduction

The Maine lobster fishery involves some of the best-documented examples of indigenous, or "folk," management practices (see Acheson 1972, 1975a, 1975b, 1987, 1988, 1989). The Maine lobster fishery is also considering a number of new formal regulations. Following James McGoodwin's suggestion that "we must incorporate indigenous means of management wherever practical and appropriate into fisheries management regimes" (McGoodwin 1990:183), this chapter addresses the question of what the practical and appropriate role of indigenous practices is in the formation of new formal regulations in the Maine lobster fishery and in fishery policy in general.

This question is approached by means of an examination of the folk practices and formal regulations of the lobster fishery along the northwest coast of Newfoundland. This approach has been chosen because this Newfoundland fishery is currently governed

This chapter draws upon information gathered during five years as a lobster fisher in Maine and two years of fieldwork in Newfoundland. The Newfoundland fieldwork was funded by a postdoctoral fellowship from the Institute of Social and Economic Research, Memorial University of Newfoundland.

by many of the formal regulations (i.e., season limits, limited entry licensing, territorial divisions, and trap limits) that are being considered for the Maine lobster fishery, and the Newfoundland fishery is generally considered to be successfully managed by both participants and regulatory officials. Therefore, an examination of the relationship between folk practices and formal regulations in this Newfoundland fishery might shed new light on potential relationships between folk practices and formal regulations in Maine.

The comparison reveals that the success of the Newfoundland regulations has little to do with their similarity to, or even compatibility with, indigenous practices. Indeed, the acceptance of formal conservative regulations in the northwest Newfoundland lobster fishery appears to be largely based on a realization of the nonconservative nature of the indigenous practices. This suggests that compatibility with indigenous practices is only one of a number of variables influencing the success of formal regulations and that such compatibility is often of less importance than other variables. For example, this chapter will argue that the conscious values and concerns revealed in indigenous practices are more important than the unintended effects of indigenous practices themselves. Hence, overly simplistic use of folk practices as models for formal regulations should be avoided, and special caution should be used when attempting to model formal regulations on folk practices that appear to have conservative effects but are not consciously intended to be conservative.

Folk Management in "Southern Harbor," Maine

Lobster fishers in the southwest Maine community I will refer to as "Southern Harbor" (see Palmer 1989, 1990a, 1990b, 1991a, 1991b, 1991c) engage in many of the folk practices Acheson has described among lobster fishers in the midcoast region of Maine (see Acheson 1988). Many of these are informal methods of limiting new entrants to their harbor. Although Southern Harbor fishers are considered "wimpy" compared to fishers in other Maine harbors because they do not resort to the mass destruction of gear or to personal violence to dissuade new users of their harbor, they do engage in more subtle tactics to make the lives of new fishers unpleasant and unprofitable in hopes that these fishers will choose a new profession. For example, one new fisher bought the boat and traps of a retiring Southern Harbor fisher during the middle of the fishing season. The retiring fisher politely showed the new fisher where his traps were set in the ocean to make it easier for the new fisher to commence fishing. The new fisher, however, was cleverly shown the same strings of traps several different times and ended up paying for nearly twice as many traps as actually existed. The new fisher was unable to overcome this economic setback and did not return the following season.

Although new fishers are usually simply ostracized, they may also be given misleading information concerning good fishing spots or the approach of storms. Although some of this misinformation, such as telling a new fisher that bricks or white doorknobs make good bait, is a form of entertainment, established fishers are aware of the economic advantages of discouraging new entrants. The difficulty encountered by new fishers is largely determined by the presence or absence of kinship ties to established fishers.

Southern Harbor fishers have also managed to maintain exclusive use of the fishing grounds around Southern Harbor by excluding fishers from neighboring harbors. When there appears to be an intentional encroachment across the territorial boundary by an established fisher, the outsider's traps are simply cut off. When a new fisher from a neighboring community is seen fishing inside of the Southern Harbor territory, that fisher is usually assumed to be ignorant of the rules. At this point, an established fisher in the neighboring harbor is informed about the violation and asked to correct the situation.

Southern Harbor also had a self-imposed trap limit during the 1970s and early 1980s. Every spring, the full-time fishers would meet to establish the limit for that season, a limit that gradually increased from 400 traps to 600 traps, and to elect two fishers to be in charge of observing fishers' traps to ensure that the limit was not exceeded. Such surveillance was possible because nearly all traps were placed onshore around the fishers' homes during the winter months.

Because the state of Maine is currently considering such new formal regulations as limited entry licensing, territorial restrictions, and trap limits, this would appear to be a perfect opportunity for indigenous practices to serve as models for formal regulations. The relationship between indigenous practices and similar formal regulations on the northwest coast of Newfoundland suggests, however, that the relation between indigenous practices and formal regulations may not be this simple.

Folk Management and Formal Regulations in Northwest Newfoundland

The lobster fishery along the northwest coast of Newfoundland is currently regulated by many of the restrictions being considered by the state of Maine. In addition to restrictions on small and egg-bearing lobsters, lobster fishers are subject to territorial restrictions, limited entry licensing, trap limits (600 per license), and season limits (May 5 to July 10). This lobster fishery stands in stark contrast to the other fisheries pursued in the region. This is because nearly all of the lobster regulations are supported by participants in the fishery. Hence, the lobster regulations have been relatively successful socially, as well as biologically, while fisheries such as the cod fishery are surrounded by controversy. This difference can be seen in the following statements made during a union meeting with area fishers:

> First Fisherman: The people managing the cod [fishery] should take lessons from those managing the lobster [fishery], because that's the only good fishery on the go.

Second Fisherman: But that's not because of the government.
The fishermen make the lobster [fishery] work because they
don't cheat on it.

The recent emphasis on folk management (see Klee 1980,
Maiolo and Orbach 1982, McCay and Acheson 1987, Pinkerton
1988, Berkes 1989, McGoodwin 1990) might lead one to expect that
both the biological and social success of these formal regulations
results from the fact that they were modeled on indigenous practices, but this is not the case.

The lobster fishery along the northwest coast of Newfoundland originated in 1873 with the establishment of the first cannery.
Harvesting and canning lobsters remained an important activity
until the 1930s, when canning was gradually replaced by the
shipment of live lobsters to Boston (see Sinclair 1985:35,42). The
enforcement of formal regulations along the northwest coast of
Newfoundland before the 1960s was even more limited than along
the rest of the Atlantic coast of Canada (see Scott and Tugwell 1981).
From the 1920s until the 1960s, when some of the current regulations started to be enforced, the formal regulation of the lobster
fishery along this isolated and sparsely settled coast fell under the
jurisdiction of one wildlife officer. This man, now in his eighties,
recalls never enforcing any regulations, or settling any disputes,
related to the lobster fishery. He summarized the regulation of this
fishery by stating, "Them fellas were on their own."

The recollections of local residents who participated in the
lobster fishery before the current regulations were implemented
confirm the lack of formal restrictions. They also make it clear that
there was little in the way of indigenous restraint. There were, for
example, no folk restrictions on participation, and people felt it was
their "natural right" to catch lobsters. Lobsters were also often
fished "from ice to ice," meaning from the time the ocean ice melted
in the spring until the ocean froze over in the fall.

Convenience did dictate that most traps were set near the home
communities of fishers, but there were no defended territories, and

some fishers seasonally migrated more than 100 miles to fish in St. John Bay, which was known as "the home of the lobster" (see Lilly 1964). Fishers living in communities along St. John Bay would also often set traps near neighboring communities when these areas had abundant lobsters. Although this practice was not popular with local residents, it was not prevented. One fisher recalled that when he was a teenager, he was the first person from his community to set traps where the men from the neighboring community caught most of their lobsters. He noted, "They didn't like it much, but what could they do?" As another fisher stated, "We had our traditional territories, but no one cared much about tradition."

There was also no attempt to limit fishing effort for conservation reasons. Although most fishers only fished 100 to 300 traps, because that was all they could conveniently make, repair, and fish, there were a few big fishers, who reportedly fished up to 1,000 traps. Instead of facing social sanctions, these big fishers were respected as hard workers. There was also little or no conservation of small or egg-bearing lobsters, as a general policy of taking everything in the trap was followed.

The crucial point is that the nonconservative nature of the indigenous practices became evident to local residents. Some residents recalled individuals who became aware of the need to conserve the resource at an early stage and began to "take care to always throw back the short and spawny ones, even before they had to," but most reported only becoming aware of the consequences of their fishing practices when severe stock depletions occurred. Stories about overexploitation and catch failures, such as the following tale about the overfishing of one local cove in the 1930s, are common: "We were only after fishing that cove for a few years when one year you couldn't even catch enough to eat. And it stayed that way the next year, too. That's when we realized it was because we had taken everything, every lobster that was there." Residents are in near-unanimous agreement that this realization of the vulnerability of the stocks under a self-regulated fishery was, and is, the major reason for the support given the current formal restrictions. As one resident stated, "We know we have to follow the rules; we've seen what happens without them."

The Role of Folk Practices in Formal Regulations

Current calls for the inclusion of folk management practices in the formation of formal regulations are usually based on the claim that folk practices have conservative effects. Indeed, it is sometimes claimed that these conservative effects, and not the conscious reasons for the practices, are of the greatest importance to fishery managers. For example, McGoodwin states that "the question of motivation is not so important for modern fisheries management as the question of a practice's potential for controlling or limiting fishing effort" (McGoodwin 1990:131). The lobster fishery along the northwest coast of Newfoundland suggests that this is not necessarily true. In this fishery, the support for the current formal regulations resulted from the fact that participants in this fishery were aware of the nonconservative nature of their folk practices and consciously wanted the resource to be conserved by formal regulations that would prevent the continuation of traditional practices.

The vast majority of lobster fishers in northwest Newfoundland are now very conscientious about releasing small and egg-bearing lobsters. Many even attempt to toss egg-bearing lobsters back into the sea in a manner that will not damage the eggs. It is also common for lobster fishers to inform authorities when they observe a fisher intentionally trying to sell small lobsters. This is in stark contrast to the region's mobile-gear cod fishery, where the illegal selling or dumping back into the sea of small, less-marketable fish is common, and fishers rarely, if ever, report violations. Both of these fisheries — cod and lobster — have a history of nonconservative folk practices. The difference between these two fisheries is that nearly all lobster fishers are convinced of the vulnerability of their prey species, whereas many mobile-gear cod fishers are not.

I suggest that a focus on the conscious intent behind indigenous practices is always necessary to determine whether the existence of apparently conservative folk practices will lead to support for new formal conservative regulations. This can be illustrated by two indigenous practices currently followed in northwest Newfoundland. Nearly all northwest Newfoundland lobster fishers observe a taboo on Sunday fishing, a taboo that was once part of the formal regulations. When a few fishers convinced the Department of Fisheries and Oceans to allow Sunday fishing, the majority of fishers banded together and used informal means to keep the taboo uniformly observed. Although abstaining from fishing on Sundays might have a conservative effect, this was not the conscious reason for the continued observance of the taboo. Most fishers did not want to fish on Sundays for religious or social reasons and did not want anyone else to fish on Sundays because they feared those who did so would take advantage of the opportunity to steal lobsters from other fishers' traps.

Northwest Newfoundland fishers also have a self-imposed restriction on setting single traps instead of the customary strings of ten traps, and fishers who have set single traps have had their traps destroyed. Although it could be argued that this practice might also have some type of conservative effect, because strings of traps cannot be set as precisely as single traps, concern over the long-term availability of lobsters for the social group is not the reason for this ban. Single traps were banned because fishers who strategically set a few single traps near the strings of other fishers were felt to be "catching other people's lobsters."

From a regulatory perspective, the importance of these two indigenous practices lies in what they imply about the conscious values and concerns of the fishers. Instead of an interest in conservation, both of these practices indicate that the fishers are concerned about their current personal catch levels. Hence, it cannot be assumed that these fishers would support new conservative formal regulations that decreased their current personal catch levels, even if these regulations resembled established folk practices.

Conclusion: The Maine Lobster Fishery Reconsidered

The relationship between folk practices and formal regulations in northwest Newfoundland suggests that caution should be taken in using the folk practices of Maine lobster fishers as models for formal regulations. The existence of indigenous practices among Maine lobster fishers regarding entrance to the fishery, territoriality, and trap limits *may* indicate acceptance of similar formal regulations, but only if these formal regulations do not conflict with the conscious reasons for the acceptance of these indigenous practices.

It is crucial not to assume that the existence of potentially conservative folk practices among Maine lobster fishers implies the existence of conscious goals that coincide with the goals of formal regulations. Formal fishery regulations are, to a large extent, aimed at preventing activities that maximize an individual's gain but decrease the benefits of the resource for some larger social group over an extended period of time. The indigenous practices of Maine lobster fishers do indicate that they have long-range "time-horizons" (Pinkerton 1988:11). In fact, the concern over the ability of sons and grandsons to enter the fishery under limited entry licensing (see Acheson 1975b) implies time-horizons that exceed the individual fisher's own lifespan. Lobster fishers are also often willing to sacrifice for the good of other individuals. This altruism, however, often fails to coincide with the group goals of formal regulations. For example, McGoodwin (1990), following Forman (1970), attempts to explain the practice of keeping the location of particularly good fishing spots secret as a means of conserving the resource for the long-term good of the group. This functional explanation does not appear to apply to the general secrecy attributed to Maine lobster fishers (see Acheson 1988). This is because

Maine lobster fishers do share valuable information, but often only with certain individuals who are either close kin or reliable recip-rocators (see Palmer 1990a, 1990b, 1991c; Acheson 1988).

Although such folk practices as the exclusion of new fishers, territorial defense, and trap limits may conserve resources for the long-term good of a group, they are typically only supported when the conscious goals of individuals are perceived to coincide with this group benefit. When indigenous practices interfere with the conscious goals of individuals, they are likely to be opposed. For example, the self-imposed Southern Harbor trap limit ended when the interests of one fisher diverged from the interests of the group. In this case a fisher insisted that his grown son be allowed to fish his own allotment of traps even though the father and son fished together on the same boat. The other fishers saw this as a violation of their right to a "fair share" and began fishing more traps. Hence, Southern Harbor has recently seen a rapid escalation of the number of traps fished by most fishers, despite the fact that the participants are well aware that this is not in the long-term interest of the group as a whole.

A failure to focus on the conscious motivations involved in indigenous practices might also produce unsuccessful formal regu-lations by causing the value Maine lobster fishers place on self-determination to be overlooked. Although Maine lobster fishers are not independent in the sense of being unconcerned with their social environment (see Acheson 1988), the relative freedom of a lobster fisher is important to many fishers. The frequent claims of "never being able to work for someone telling me what to do all day" are often more than mere rhetoric. This means that even regulations modeled on existing folk practices may be opposed if fishers are no longer involved in the implementation and enforce-ment of these regulations (see also Andersen 1987:341). For exam-ple, despite the existence of general support for trap limits (see Acheson 1975, 1988), some of the fishers in a harbor who were required to take part in an externally imposed trap limit experi-ment in the late 1980s painted a hammer and sickle on their boats in protest of what they perceived to be a violation of their freedom and natural rights.

In a review of McGoodwin's book *Crisis in the World's Fisheries,* Peter Sinclair states that it is still "not really clear what fisheries managers have to learn from indigenous fishers' own practices" (Sinclair 1991:123). I suggest that indigenous practices can be a valuable asset to those implementing formal regulations, but not necessarily as models for formal regulations. The social acceptance of formal regulations, with all it entails for reduced enforcement costs and the conservation of the resource, depends upon the conscious concerns and values of fishers. A major value of indigenous practices lies in what they reveal about these concerns and values.

Lessons for Modern Fisheries Management

1. Some folk practices have nonconservative functions, and these should be identified by managers who hope to use folk management knowledge as models for formal regulations.
2. To achieve maximum support from fishers for proposed formal regulations, an understanding of the conscious intent behind their folk management practices should be examined.
3. When folk management practices interfere with the conscious goals of individuals, opposition and conflict are likely outcomes.
4. In some fisheries, folk management is best utilized as a way of understanding the values and concerns of fishers, and not necessarily as an appropriate template for the enactment of formal regulations.

References

Acheson, J. M.
 1972   "The Territories of the Lobstermen." *Natural History* 81(4):60–69.

1975a "The Lobster Fiefs: Economic and Ecological Effects of Territoriality in the Maine Lobster Industry." *Human Ecology* 3(3):183–207.

1975b "Fisheries Management and Social Context: The Case of the Maine Lobster Fishery." *Transactions of the American Fisheries Society* 104(4):653–668.

1987 "The Lobster Fiefs Revisited: Economic and Ecological Effects of Territoriality in Maine Lobster Fishing." In *The Question of the Commons: The Culture and Ecology of Communal Resources,* ed. by B. J. McCay and J. M. Acheson. Pp. 37–65. University of Arizona Press: Tucson.

1988 *The Lobster Gangs of Maine.* University of New England Press: Hanover, New Hampshire.

1989 "Where Have All the Exploiters Gone? Co-Management of the Maine Lobster Industry." In *Common Property Resources: Ecology and Community-Based Sustainable Development,* ed. by F. Berkes. Pp. 199–217. Belhaven Press: London.

Andersen, E. N., Jr.
1987 "A Malaysian Tragedy of the Commons." In *The Question of the Commons: The Culture and Ecology of Communal Resources,* ed. by B. J. McCay and J. M. Acheson. Pp. 327–341. University of Arizona Press: Tucson.

Berkes, F., ed.
1989 *Common Property Resources: Ecology and Community-Based Sustainable Development.* Belhaven Press: London.

Forman, S.
1970 *The Raft Fishermen.* University of Indiana Press: Bloomington.

Klee, G. A., ed.
1980 *World Systems of Traditional Resource Management.* John Wiley and Sons: New York.

Lilly, H. D.
1964 "Marine Environment Lobster Survey: St. John and Pistolet Bay, Newfoundland." Department of Fisheries and Oceans: Ottawa, Canada.

Maiolo, J. R., and M. K. Orbach, eds.
1982 *Modernization and Marine Fisheries Policy.* Ann Arbor Science Publishers: Ann Arbor, Michigan.

McCay, B. J., and J. M. Acheson, eds.
1987 *The Question of the Commons: The Culture and Ecology of Communal Resources.* University of Arizona Press: Tucson.

McGoodwin, J. R.
1990 *Crisis in the World's Fisheries: People, Problems, and Policies.* Stanford University Press: Stanford, California.

Palmer, C. T.
1989 "The Ritual-Taboos of Fishermen: An Alternative Explanation." *Maritime Anthropological Studies* 2(1):59–68.

1990a "Telling the Truth (up to a Point): Radio Communication Among Maine Lobstermen." *Human Organization* 49(2):157–163.

1990b "Balancing Competition and Cooperation: Verbal Etiquette Among Maine Lobstermen." *Maritime Anthropological Studies* 3(1):87–105.

1991a "Organizing the Coast: Information and Misinformation During the 1989 Maine Lobstermen's Tie-Up." *Human Organization* 50(2):194–202.

1991b "The Life and Death of a Small-Scale Fishery: Surf Clam Dredging in Southern Maine." *Maritime Anthropological Studies* 4(1):56–72.

1991c "Kin-Selection, Reciprocal Altruism, and Information Sharing Among Maine Lobstermen." *Ethology and Sociobiology* 12:221–235.

Pinkerton, E., ed.

1989 *Co-Operative Management of Local Fisheries.* University of British Columbia Press: Vancouver.

Scott, A., and M. Tugwell

1981 "The Public Regulation of Commercial Fisheries in Canada: The Maritime Lobster Fishery." Technical Report Series, Economic Council of Canada: Ottawa.

Sinclair, Peter R.

1985 *From Traps to Draggers.* Institute of Social and Economic Research, Memorial University of Newfoundland: St. John's.

1991 "Review of *Crisis in the World's Fisheries* by J. R. McGoodwin." *Maritime Anthropological Studies* 4(2):123.

10

Two Tales of a Fish: The Social Construction of Indigenous Knowledge Among Atlantic Canadian Salmon Fishers

Lawrence F. Felt

There's as many fish as there ever was. It's just anglers talking.
Commercial salmon fisherman on Newfoundland south coast.
November 21, 1989.

No b'y, they're gone. Too many been caught, especially the large mother fish.
Commercial salmon fisherman on Newfoundland northeast coast.
November 22, 1989.

Increasingly, meaningful participation of fishers, particularly localized shore-based ones, has been promoted as essential if the pervasive experiences of dramatic depletion among marine resources are to be reversed. Several reasons have been offered for the inclusion of fishers in resource management; an important one has been that harvesters possess extensive knowledge of the species pursued and the wider marine environment, knowledge that is useful for sound management (Berkes 1977, 1987; Cordell 1974; Kloppenberg 1991; McGoodwin 1990; McKay 1981, 1987; and Morrill 1967). Though typically embedded in a culture and cosmology alien to the natural scientists housed within state bureaucracies, such indigenous, or "folk," knowledge may at the very least serve as an important, useful supplement to more conventional scientific information and at most provide the basis for an alternative understanding of ecosystem operation (Kloppenberg 1991).[1] Nakashima (1990) makes this case in his study of Hudson Bay Inuit hunters' knowledge of eider duck populations in the path of oil exploration. "We must be careful not to lose perspective. Scientific method, quantification and statistical significance are ideals. . . . Shortcomings in our scientific knowledge are staggering. . . . Traditional environmental knowledge presents us with a rare opportunity which should not be ignored" (1990:23). Nakashima then demonstrates the complex knowledge Inuit hunters possess of eider duck spatial and temporal distributions as well as their interrelationships to the wider marine environment. Although communicated in stories and anecdotes rather than frequency distributions and confidence limits, Inuit knowledge far outstrips information available in current scientific literature in understanding the eider's winter ecology. As such, Nakashima argues Inuit knowledge is invaluable as a baseline for assessing any impacts of petroleum exploration and development.

Without denying the general validity of Nakashima's (and others') research, which argues that isolated peoples relying heavily on a marine resource for subsistence develop intimate knowledge of it, it is also useful to flesh out the utility, even the accuracy, of such knowledge for more general marine management. For instance, when fishing is undertaken for external markets and harvesting is circumscribed in various ways by state regulations, the case for incorporating localized fishers' knowledge may be problematic (Berkes 1987).

Consider the two initial quotes from Newfoundland commercial salmon fishers who, utilizing identical gear and fishing stocks that overlap to a significant extent, reach dramatically different conclusions about the resource's health.[2] Considerable evidence suggests that each view is not idiosyncratic, but instead is shared by a considerable number of fishers. How can such a divergence of opinion be explained? More to the point for proponents of cooperative management, what are the implications for utilizing local knowledge for resource management when conclusions drawn from that knowledge may be inconsistent, even contradictory?

In this chapter I shall argue that lack of consensus, and even contradictions, among fishers need not negate the relevance and utility of their knowledge for resource management. The key to successful utilization, however, lies in first understanding the processes and context within which local knowledge is produced. In contemporary language, the knowledge must be deconstructed and then reconstructed (Pinch 1986) for potential management use. Fishers' knowledge, in other words, must be primarily understood as a social construction in which particular experiences are given meaning within a specific cultural context. As such, the processes whereby something comes to be known assume at least as much importance as the knowledge claim itself. Moreover, I feel that where local knowledge is highly fragmented (nonholistic) and largely instrumental (practical) and where fishing strategies are highly regulated by external bureaucracies, conclusions derived from local knowledge will be highly colored by such external factors as imposed management restrictions, competition with other claimants for the resource, and degree of participation in wider fisher organizations battling other claimants to the resource. Thus, fishers' assessments of resource health must be seen as more than mere extrapolations from their general, localized knowledge. Failure to understand this complex process of knowledge construction may lead to the unfortunate decision to simply discount fishers' assessments that differ from "scientific" ones as misguided, ignorant, or worse. Such a conclusion undermines the general claim to legitimacy for indigenous knowledge, as well as precludes potentially insightful understandings about a resource, simply because the general conclusion appears at odds with other, dominant views.

The Problem

The Atlantic salmon *(Salmo salar)* is an anadromous fish inhabiting the North Atlantic Ocean. Highly migratory during its adult life, Canada's Atlantic salmon journey from streams in eastern Canada as far east as southwest Greenland, then return to natal rivers to reproduce. Due to its high economic value, the Atlantic salmon is harvested throughout its migration. Unlike Pacific salmon, Atlantic salmon survive after spawning and, if the many perils are successfully avoided, may return to the same section of river or stream to reproduce a second or third time, or even more. At this time the largest specimens may weigh upwards of fifty pounds. In the most easterly Canadian province, Newfoundland and Labrador, approximately 3,000 inshore fishers harvest these fish utilizing small boats and fixed gill nets. The prime harvesting periods range from late May to the end of June for the island portion of Newfoundland and from late May through July for mainland Labrador.

For management purposes, the province is subdivided into fourteen zones beginning in northern Labrador and moving clockwise along the east, south, and west coasts (see Figure 10.1). Because reference is frequently made to these salmon management areas, it will be useful for the reader to become familiar with the zones depicted in Figure 10.1.

Salmon represent between 7 and 18 percent of the landed value for most commercial fishers in the province of Newfoundland and Labrador, though this may increase by an additional eight to ten percentage points in any given year, depending upon the abundance of other marine resources such as cod, lobster, capelin, and the other species pursued (Pinfold 1990). For the Labrador zones 1 and 2, plus island zones 3 and 11, the percentage may go considerably higher, sometimes accounting for up to 60 percent of annual earnings. The salmon's importance to local fishers is also increased considerably depending on the timing of their return. Though current Canadian government annual unemployment

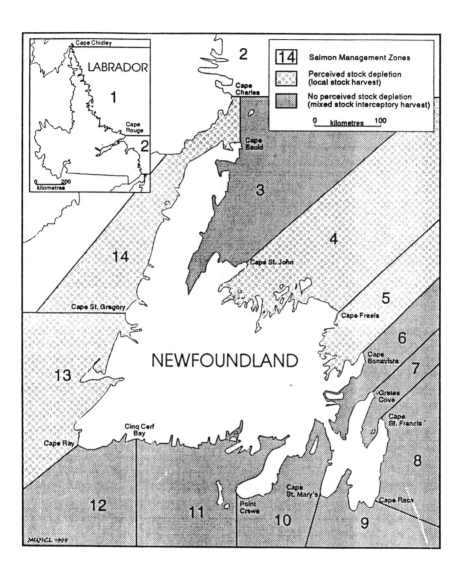

Figure 10.1 Salmon management zones in Newfoundland.

insurance payments terminate on May 15, cod and other valuable fish are typically not available until late June or July. Thus, salmon landings often bridge the gap between the last unemployment insurance checks and the first fisheries earnings of the new year. This importance has been heightened somewhat in recent years with significant curtailment in the winter seal fishery, which performed a similar function for many inshore fishers.

Gill nets utilized to catch salmon average fifty fathoms (300 feet) in length and have five-inch-square mesh. Current regulations allow for up to 200 fathoms of net. Typically, nets are hung between floats to fish in the top three fathoms of water. Most nets are located within a quarter mile of land, typically off a jutting headland, although fishers in zone 11 often set nets up to two miles from land.

Throughout the last decade considerable controversy has surrounded the salmon fishery. Commercial landings have declined significantly from approximately 1,500 metric tons per year in the 1970s to just over 400 in 1991 (DFO 1992). The decline has been particularly precipitous since 1988. Highly organized recreational groups have mounted political campaigns calling for significant reductions, if not outright elimination, of commercial fishing in order to rebuild the stocks. Many commercial fishers, through their union, have countered by denying salmon stocks are declining and have branded recreational groups as greedy and insensitive to the important niche commercial fishing plays in the survival of inshore fishers, their communities, and their economies. In the fishers' view, the declines in commercial harvests are explained as a result of increasingly restrictive management changes that significantly curtail fishers' efforts. For their part, state-employed biologists have been equivocal in their pronouncements, avoiding until very recently any official comment on the health of salmon stocks. Only since 1990 have they been willing to publicly reveal what are called spawning estimates, that is, a requisite number of female fish and eggs for full production in various provincial rivers. Using these estimates, the majority of Newfoundland rivers appear to be producing at between 30 and 60 percent of capacity (DFO 1992).

In this highly politicized context, one might expect that those wanting to curtail another user would claim serious stock decline, and, reciprocally, those implicated as a major cause would deny it. Indeed, at the level of media exchanges and formal representations to government, such a conclusion appears adequate. Closer examination, however, reveals significant divisions within both recreational and commercial groups. The division among fishers, for example, which is reflected in the two opening quotes, appears to have a clear geographical basis. For example, most fishers expressing concern for overexploitation and acknowledging considerably reduced salmon populations reside in management zones 4, 5, and 13, whereas those claiming that the allegations are merely a self-perpetuating recreational mythology dwell in zones 3, 7, 8, 10, and 11 (See Figure 10.1). Although this geographical division generally corresponds to fishers' dependence upon the resource, it is clear that their responses are considerably more than defensive rationalizations. Indeed, advocates of each position are able to muster a considerable amount of resource knowledge to substantiate their respective claims.

Setting and Data

The analysis in this chapter is based on various forms of data collected over the past six years from salmon fishers living in salmon management zones 3, 4, 5, 7, 8, 10, 11, and 13. As indicated earlier, generally speaking, fishers from the northeast and southwest coasts of Newfoundland (zones 4, 5, and 13) perceive a serious stock decline, while those on the southeast and south coasts (zones 3, 7, 8, 10, and 11) do not. Before attempting to explain this difference in perception, a brief overview of salmon fishers and their communities is in order.

Superficially, salmon fishers and their communities appear highly similar. With only a handful of exceptions, they utilize small, open boats twenty-five feet in length or less and gill nets of identical mesh size. In addition, all share similar management restrictions on gear amounts and on season. Finally, all are pluralistic fishers possessing multiple licenses and pursuing several different species, such as lobster and cod, throughout the year. Other than their possession of a commercial salmon license, there is little to differentiate them from the approximately 15,000 other inshore fishers scattered around Newfoundland's coast.

The overwhelming majority reside in small, rural communities called *outports.* Whereas an earlier generation of such outports owed their entire existence to the inshore fishery and its associated saltfish processing (see Faris 1972 and Firestone 1967), economic diversification and technological changes in the fishery leading to deep-sea harvesting and frozen fish processing have lessened somewhat the dependence of such communities on this particular pursuit. Nonetheless, the inshore fishery remains economically important to a large number of families within the communities, with between 15 and 30 percent of community employment directly dependent upon it (DFO 1991). Even though actual dependency upon a traditional inshore fishery has lessened, the cultural values and norms associated with such a fishery continue to inform residents' understanding of the world around them and continue to condition their interaction with fellow residents and outsiders alike (Palmer 1992; Porter 1983, 1985).

To collect information, semistructured interviews were completed with seventy-one commercial salmon fishers. The interviews were typically done on stages or in fishers' homes. A similar format was utilized in which opinions were gathered on the status of the stocks, reasons for any decline, indicators fishers used to detect decline, and strategies necessary to rebuild stocks if they were thought to be in decline. The interviews invariably expanded to include a wide range of subjects and opinions ranging from changing lifestyles in the communities to frustrations with government. On many occasions, interviews occurred in the form of a directed discussion between the researcher and two or three fishers simultaneously. This approach was utilized particularly when fishers were returning from hauling nets. The majority of the interviews were completed between November 1989 and November 1991.

In addition to interviews, observations were carried out at several resource workshops convened by the Canadian national and Newfoundland provincial governments between December 1984 and March 1992. These workshops included one sponsored by the Atlantic Salmon Advisory Board (ASAB), the Canadian government's primary consulting mechanism for users of the salmon resource, as well as numerous less formal workshops on salmon management. Government materials including reports, media releases, and scientific assessments were made available to the researcher. In addition, interviews were completed with union personnel representing salmon fishers, as well as with representatives of recreational salmon fishing associations.

Forms and Context of Fishers' Knowledge

Fishers' judgement of a resource's health begins with their more general knowledge of that resource and its larger marine environment. Interviews suggest some similarities but also important differences in understanding between those acknowledging stock decline and those dismissing the claim. These differences appear largely attributable, as argued in the next section, to the geographic proximity of salmon-producing rivers to fishers' communities and to the implications of such proximity for the content of local understanding.

As reported by Berkes (1987), Nakashima (1990), and others, fishers communicate their understanding about their world through personal anecdotes and long stories or yarns about community members of yesteryear. Humor is frequently used, as are "cuffers," or exaggerated stories told in competition with other fishers. A typical example comes from Art, a salmon fisher from

Fogo Island (zone 4), in which he recalls how his grandfather discovered how to set a salmon net in relation to an iceberg:

> Grandfather was out in early June, and it was some hot. He'd got some salmon and was worried about how he'd keep them fresh. There were some [ice]bergs nearby, so he thought he'd get some ice. He drawed his boat up alongside and chopped some ice loose. The ice fell into the water, so grandfather dipped it out with a dip net he had in the boat. Out came five salmon. He looked down and saw hundreds. . . . Well, after that grandfather always set his nets anchored to a berg or at least near it, and he got more fish than anybody else. . . . He never told many about it, but he sure caught the salmon for awhile. But I guess the fish got smart, 'cause it doesn't work as well nowadays.

One further consideration is important before comparing and contrasting fishers' knowledge about salmon. Regardless of the extent and content of knowledge that emerged from interviews, this knowledge was always framed within a larger context of fatalism, doubt, and unpredictability. That is, no matter how much fishers might know, they would invariably begin and conclude their narratives with a self-deprecating disclaimer that the ocean was a mysterious, dangerous, and unpredictable master revealing little of its mystery to those attempting to survive by harvesting its resources. In the words of one fisher:

> It's [fishing] a hard life. You never knows what's going to happen out there. I been at it thirty-five years, and I still get surprised every time I go out. Salmon have always been up and down. You know, plentiful for awhile and then scarce as anything. They're down now, for sure, but they could bounce back, I guess. If they're there, I can fish pretty well, but nobody knows for sure whether they're going to be there or not.

Sider (1989) has examined the larger outport cultural configuration in which danger, powerlessness, and fatalism are central

themes when confronting the natural (as well as external economic) environment. Although a comprehensive statement of how
such self-effacement arises and is reproduced takes us far beyond
the more modest objective of this chapter, one important consideration for understanding fisher knowledge should be mentioned.
Fishers do not easily offer conclusions from their experiences and
are reluctant to claim great knowledge, even when the extent of
their understanding is substantial. Without minimizing the sincerity
of fisher disclaimers, it is equally important to understand that they
are part of a protective cultural adaptation to essentially uncontrollable physical and social environments. Not to understand this
runs the risk of confusing cultural rhetoric with ignorance.

Two Tales of a Fish: Variability in Fisher Knowledge

The two contrasting assessments of salmon stock health are
drawn, in part, from two very different understandings of the
resource. Each understanding can be summarized in terms of what
might be called a knowledge profile that captures the essential
elements of each type.

Profile 1

Fishers in management zones 3, 7, 8, 10, and 11 possess what
is best described as an instrumental knowledge of salmon. This
knowledge contrasts with the characterization of small-scale fishers
as "ecosystem people" (Dasmann 1974, Klee 1980) who dwell
within a single, or at most a limited number of interrelated,
ecosystems and who have a holistic understanding of the various
natural and animal elements within it. While encompassing detailed information on salmon behavior, instrumental knowledge
is directly linked to harvesting. Salmon movement according to
wind, tide, ice conditions, time of year and day, as well as seawater
temperature is extensively well-known, with a highly colorful

local vocabulary to differentiate the myriad combinations of inter-action between these physical elements. Conversely, there is little understanding of the fish's life cycle, feeding habits, migration patterns beyond the immediate area, or relationship to other ele-ments of the wider marine environment.

Less than a third of the interviewees from these zones knew that salmon were anadromous and ascended rivers to spawn. Though all distinguished between early-arriving larger salmon and the trailing smaller grilse, or *jacks,* other variations in size and shape were not recognized to any great extent. Because the arrival of salmon in these zones follows that of the capelin, a small herring-like pelagic fish, most fishers assumed that salmon must be follow-ing them for food. Where salmon had been and where they were heading as they passed by fishers' berths remained unknown, though many had heard biologists say that the bigger fish had been feeding somewhere off Greenland. And, in my interviews of thirty-nine fishers from these zones, none claimed that the resource was in any serious trouble.

Before attempting to account for this particular kind of knowl-edge, it is important to compare and contrast it to another profile. This profile characterizes those fishers professing some degree of concern about declines in salmon populations.

Profile 2

Salmon fishers in zones 4, 5, and 13 also possessed extensive instrumental knowledge on how to catch salmon. It much more closely approximates the "ecosystem" orientation referred to ear-lier. Several examples may serve to illustrate. In addition to distin-guishing jack fish from large salmon, most of this second group described certain physical characteristics such as size of head in relation to the rest of the body, the shape of the head, or the shape of the body, characteristics that they claimed were associated with particular rivers. A fisher from zone 4 drew the outlines of several different shapes of fish on a brown paper bag during an interview and informed me as follows:

> See, this [drawing on a brown paper bag] is how Ragged Harbour fish look compared to Gander fish. The Ragged Harbour fish is more slinky. Don't know why they's different, but guaranteed you can tell the difference. Maybe it's due to the lack of feed in the river. I know young salmon eat insects, and you just don't see as many black flies and such over on Ragged Harbour.

Wider natural relationships, including those linking inland climate with salmon production, were also alleged. Several fishers, for example, related the lower catches in 1990 and 1991 to an exceptionally dry summer of 1987, in which adjacent riverbeds nearly dried up and large numbers of immature salmon died.

Fishers from these zones also frequently compared fish in terms of flesh color. Dark pink, nearly red flesh was associated with consuming large quantities of shrimp and hence migration north either off known Labrador shrimp areas or off Greenland. Several fishers recounted stories of how they came to link Greenland with Newfoundland salmon. According to one fisher:

> You hear people talking about the Greenlanders catching our salmon on Fishermen's Broadcast [a popular CBC radio program on the fishery out of St. John's], but I never understood why, or even where Greenland was, really. I was in town [St. John's] to a union meeting once about ten year ago, and there were some folk working on a Danish ship fishing for shrimp off Labrador. Anyways, they [the Danes] were teaching our boys how they fished for shrimp off Greenland. It all made sense to me then. That's where some of the salmon went. . . . Since it was so far away, it probably meant that the biggest, reddest ones went there, or something like that.

Many fishers also questioned the use of highly selective gill nets. One result, it was argued, was a decline in the size of fish caught over the last fifteen or twenty years. One fisher likened the situation to farmers who continually slaughter their best cows, leaving only the runts as breeders.

Explanations for Variability in Fisher Knowledge

How do the two groups of fishers come to acquire such different knowledge about Atlantic salmon? The critical factor appears to be the geographical proximity of salmon-producing rivers to fishers' communities. Consider Profile 1 fishers resident in salmon management zones 3, 7, 8, 10, and 11. With the exception of a few rivers within zone 10, these parts of the Newfoundland coast possess no major salmon-producing rivers. Thus, salmon harvested by fishers resident in these zones are intercepted as they return to other parts of the province or to mainland Canada. Figure 10.2 illustrates salmon migration routes. Note how the three major migration routes taken by Atlantic salmon around the province are drawn.

According to government tagging studies, salmon typically migrate very close to shore, often within a few hundred yards of it, in one of three general patterns around the island of Newfoundland. The first movement involves fish returning to Labrador and to Quebec's north shore. Upon returning from the North Atlantic, the salmon swim southwesterly into management zone 3 and then northward to either ascend southern Labrador rivers or migrate further westward into Quebec rivers in the Gulf of St. Lawrence. The second pattern involves fish returning to zones 4 and 5. These fish return to Newfoundland near the boundary between zones 3 and 4, from where they move in an easterly, clockwise direction until they find their natal watershed among the many that line this part of the island. The third migration path involves a southwest movement from the Labrador Sea or western Greenland to Newfoundland shores, then a clockwise movement around the island. Upon "striking" the Newfoundland coast, they migrate through

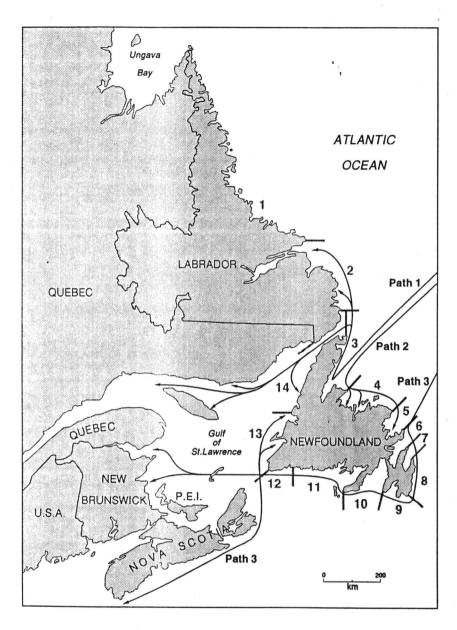

Figure 10.2 Salmon migration paths throughout Atlantic Canada.

zones 6 to 12 and thence into the Gulf of St. Lawrence. From there, various stocks separate, some moving to the west coast of New-foundland (mostly to zone 13) while others continue to mainland watersheds in the Canadian provinces of New Brunswick and Quebec.

The implications for fishers located in each of the migration routes are quite different. For those in zones 6, 7, 8, 10, and 11, salmon are available to harvest for only a short time each year — typically two to three weeks. As the gear utilized is fixed, once fish have migrated past, fishers can only hope to catch a few stragglers or a small amount of locally produced fish. Because other species become available at approximately the time the migration ends, effort is typically redirected to other species. As one south coast salmon fisher remarked, literally, of his window of opportunity:

> I looks out my window here [pointing to kitchen window], and sometimes if the water is calm, you can see a few jumping. By the first week of June they's mostly gone. Since boys from Rencontre [a nearby community] set their nets down thataway [points west beyond the view from the window], I only got a fairly small area here and two or three weeks to catch them. Then they're gone.

A second consequence of this fleeting relationship with salmon is that fishers possess limited knowledge of the fish's life cycle. Salmon are encountered only as adults migrating past fish-ers' communities. As indicated earlier, it was quite common to believe that salmon spawn at sea like mackerel, herring, and other pelagic species. Due to this belief, the link between terrestrial changes, that is, warm, dry summers with severely reduced flow in rivers, and subsequent numbers of adult salmon in the sea was not often clear. A third consequence, and one more directly related to fishers' assessment of the resource, was that the numbers of fish migrating past in relation to the numbers harvested was so large that a significant reduction in total numbers might very well go unnoticed. To place this in perspective, as many as 750,000 salmon typically might migrate across the southern coast of Newfoundland. Three hundred and fifty fishers might catch roughly 35,000 fish, or less than 5 percent of the total.[3] Even if total migration were reduced by 50 percent, this would not necessarily lead to a significant reduction in harvest.[4] As long as effort remains similar and the fish are not unusually late due to ice conditions, harvest levels are likely to be fairly stable over time, despite substantial variation in the numbers migrating by. An examination of catch statistics over the last thirty years for these zones supports this conclusion. Of all fourteen salmon management zones for the province, zones 7, 8, 10, and 11 have demonstrated the greatest stability in harvest levels. Recognizing any decline in numbers of salmon is exceedingly difficult in such a situation.

Fishers in zone 3 have a similar fishery, although fish are typically available for an extra week or two in comparison to the southeast and south coasts. Salmon fishers in zones 4, 5, and 13 experience a different situation. Although there is some intermixing of nonlocal and locally produced fish in these zones, harvesting is directed much more toward locally produced fish.[5] Because a majority of the important salmon rivers of the province are located within these zones, they have historically been, along with Labrador, the most productive areas of the province. These fishers also enjoy a longer season. This is primarily a result of timing differences in the salmon returns to local rivers. It is quite common for two rivers within a few miles of each other to have the salmon arrival occur two to three weeks apart. A result is that if other species such as cod are in limited supply, salmon fishing may be continued into the middle of August (though this is the exception rather than the rule). The geographical proximity of salmon river and fishers allows for a more comprehensive, integrated understanding of salmon. Rather than a fleeting encounter as they pass by, fishers encounter the salmon at various stages of their life cycle. Thus, although it may be understandable that many fishers in zones 3, 7, 8, 10, and 11 would not know that salmon are anadromous or reproduce in fresh water despite their adulthood in the ocean, fishers in zones 4, 5, and 13 frequently entrap spawned-out salmon, or *kelts,* returning to sea after overwintering in the river. These kelts are caught in nets set for cod and other fish. Similarly, small-mesh traps utilized for catching capelin, a small, smeltlike fish, often unintentionally entrap large numbers of silvery *smolt,* or juvenile Atlantic salmon embarking on their maiden voyage to the sea.

Fishers in these zones also associate certain physical charac-teristics of smolt and kelts with specific river systems. They are also sensitive to the numbers of smolt and kelts they encounter as an indicator of future returns. Linking this knowledge to certain climatic changes allows for some predictions of how numerous fish will be in succeeding seasons. The following discussion with a fisher from zone 4 captures this integrated, "ecosystem" type of understanding:

> I was going backcountry to do some berry picking with the missus. It was August of 1987. Anyway, I noticed on Deadwolf Brook that the water was lower than I'd ever seen it before. The little salmon peels [parr] would be balled up where a little spring or something ran into the brook. I said to myself that if we don't get some rain, they're going to have problems. Sad to say we didn't get no rain for a couple more weeks. I noticed next spring that there were fewer spawned-out fish, and I didn't get as many silvery ones [smolts] in the capelin traps. . . . Expect they'll be fewer salmon in a few years.

Where catches depend heavily upon local rivers, harvests often represent a significant proportion of total local production. Canadian government research suggests, for example, that more than 65 percent of locally produced small salmon and as much as 90 percent of larger, multi-sea winter salmon may be captured in gill nets.[6] Given such high exploitation rates, a change in the total abundance of fish is quickly experienced, unlike in those zones harvesting fish that are mostly migratory.

Thus far, I have tried to show that salmon fishers in different regions of Newfoundland possess quite different understandings of the resource. These differences in themselves, however, are hardly adequate to comprehend the adamancy with which fishers in zones 3, 7, 8, 10, and 11 reject any suggestion of stock decline and characterize changes in salmon management as a thinly disguised plot to reallocate catches to recreational anglers. Consider this comment, which is quite typical of a number recorded from fishers in zones 10 and 11: "It's a damn plot! Those 'bloods of bitches' [anglers] just want the fish for themselves. It's just that they got the damn politicians to listen to them 'cause they got lots of money and can write fancy papers. They're [the government] destroying our way of life and taking food from my children."

The explanation for such adamancy and denial of any stock decline amongst these fishers lies in the relationship between their partial, instrumental knowledge, on the one hand, and the consequences of certain important external factors on the other. Two general forms of external factors are critical: (1) changes in state fishery regulations and (2) participation in the union representing fishers in the province. In combination with an instrumental understanding, they lead fishers to reject claims of stock decline and to perceive salmon management as almost exclusively a political battle. For fishers possessing the more holistic view presented in Profile 2, these factors either have minimal consequence or are absent completely. This statement requires some elaboration.

State Regulatory Changes and Fisher Perceptions of Abundance

As a result of heavy lobbying from angler organizations and provincial governments in the mainland Canadian provinces of New Brunswick, Quebec, and Nova Scotia, the Canadian government initiated a number of significant regulatory changes commencing in 1984 in the commercial Atlantic salmon fishery. The major objective of such changes was to reduce interception of mainland-origin salmon by Newfoundland fishers. Among these changes were the removal of part-time fishers, limiting gear to 200 fathoms of gill nets, delaying the opening of the season until June 5, and, after 1990, the imposition of commercial quotas for each management zone.

The effects of these changes have been felt most noticeably in zones 7, 8, 10, and 11. The delaying of the opening of the fishing season to June 5 means that in a year free of ice and unusually cold seawater temperatures, a large proportion of the stocks moving along the east and south coasts will be clear of fishers' nets before the season opens. Even when the fish are delayed, the twenty-five-metric-ton quotas for zones 10 and 11 (with much smaller quotas for zones 6, 7, and 8) represent more than a 50 percent reduction from pre-1984 levels. It is of little surprise that many fishers feel the fishery has been taken away from them. As one fisher laments:

> She's gone, b'y. They took her away from us. We used to have two or three weeks of good fishing. Now you got maybe a week, depending upon ice. It's like a race now with the quota. You gotta get out there and get 'em quick before they closes the thing down. What the hell is 25 ton anyway? It's crazy. It's all because of those mainland angler groups. I don't see any signs that there are fewer salmon nowadays. Of course, the way we're regulated now, I guess you couldn't tell even if there were.

The above quote captures a number of widely felt sentiments. Regulatory changes have severely reduced these fishers' profits from this fishery. Moreover, the quotas have eliminated any possibility, however remote, that the fishers could detect any reduction in stocks. It is therefore easy to see why the state authorities' attempts to legitimate such management changes in the name of conservation produce widespread skepticism and cynicism.

The effects of regulatory changes have been much less noticeable in zones 4, 5, and 13. Due to ice cover and the timing of salmon returns, fishing seldom begins until June 10 in zone 13 and the beginning of July in zones 4 and 5. In addition, the quotas, though reduced from historic levels, do not represent anywhere near as drastic a curtailment as they do for the other zones, and thus the regulatory changes do not interfere between fishers and their understanding of the resource. In the words of a zone 13 fisher:

> Oh, they [regulatory changes] have made some difference. A lot depends upon the ice conditions. Generally, however, I think we've benefitted more than been hurt by them. The small quotas on the south coast has meant that more of our fish as well as that of the mainlanders has gotten back. I knows we been catching more larger fish than we used to. Really all that the government did is take away fish from the south coast and give them to us. I'm not sure it did any good for conservation or that sort of thing. It ain't what it used to be. They ain't enough getting into the rivers, and poachers are getting most of the ones that do get there.

Unlike in the other zones, regulatory changes have not contributed to a more jaundiced, political interpretation of motives in zones 4, 5, and 13. As indicated above, one reason for this is that these changes have not undermined fishers' link to the resource. As the above-quoted fisher clearly indicates, stocks have continued to decline in nearby rivers despite (and perhaps even because of) the regulatory changes.

Although not a regulatory issue, an additional feature of bureaucratic management that has contributed to sustaining these rather divergent views of resource health is the scientific assessment process conducted by fisheries biologists. Theoretically, biological assessment plays a critical role in management decisions about harvest levels. Biologists on the Canadian mainland monitoring important salmon rivers have been nearly unanimous in expressing their concern about low numbers of salmon in mainland rivers. Their cries have been an important legitimating factor in the mainland anglers' and government's lobbying for commercial cutbacks.

In Newfoundland, biological assessments have been very general and vague on critical issues. Provincial assessments are undertaken from two locations. Zones 1 through 12 are assessed from a regional office in the capital city of St. John's. The assessment staff consists of perhaps fifteen biologists, technicians, and support staff. Zones 13 and 14 are assessed from Moncton, New Brunswick, because the western part of Newfoundland along with parts of the Canadian provinces of New Brunswick, Prince Edward Island, and Nova Scotia have been consolidated into what is known as the Gulf of St. Lawrence Management Zone (or Gulf Zone, for short). Until the last few years, assessment has consisted largely of monitoring commercial and recreational catches for long-term fluctuations and monitoring a series of index rivers around the island with either fishways or counting fences or both. At these sites salmon are counted and measured as they move upstream from the sea, and frequently juvenile salmon, or smolt, are counted as they swim downstream to their first contact with salt water. Index rivers are used as rough indicators for the regions surrounding them. Over the last ten years, index rivers have been established in zones 4, 5, 9, 10, 13, and 14.

Perhaps the most crucial questions to be asked of resource management are (1) What is the production capacity of a resource, assuming habitat is maintained? and 2) Given the answer to the first question, are current harvesting levels sustainable? During the last thirty years, mainland biologists have undertaken extensive analyses of individual river systems. These analyses include morphological mapping of rivers, juvenile population estimating, and enumeration of smolt heading out to sea. For at least the past twenty years, such assessments have led them to conclude that the vast majority of mainland rivers were alarmingly below capacity. Newfoundland assessments, in contrast, had traditionally eschewed any estimating of individual river productivity and had instead consisted of recording commercial and recreational catches as well as adult returns on a limited number of rivers. These totals were then arrayed with previous counts. The consensus of biologists from the Newfoundland region was that trends of all three data sets were stable over the last ten years.

Such stability, in their estimation, suggested that Newfoundland stocks were holding their own and that any additional decline in provincial stocks should be attributed to factors other than the Newfoundland commercial fishery. Some concern was expressed about an apparent decline in numbers of larger salmon, but as it was alleged that most Newfoundland rivers produced smaller, one-sea winter grilse anyway, the decline was difficult to interpret in light of recent regulatory changes in the commercial fishery. When pressed for more specific information similar to that generated in other Canadian provinces, biologists explained that the resources they needed to undertake such assessments were unavailable. These pronouncements did little to lessen the conflict amongst various users. Essentially, each constituency attached its own interpretation to the data that had been presented. Depending upon one's position, either the stocks were holding their own and therefore not in trouble, or they had been severely depleted from earlier levels and were stable at a fraction of their potential number.

After considerable public pressure, Newfoundland biologists agreed to provide estimates of current production for a limited number of their index rivers, beginning in 1990. Three were analyzed in greater detail: Conne River in zone 10, Gander River in zone 4, and Fischel's Brook in zone 13, all of which were found to vary considerably in their current production levels. The Gander River was estimated to be producing at approximately 25 to 30 percent, Fischel's Brook at 20 percent, and the Conne River at 104 percent. The explanation offered for the seemingly anomalous Conne River situation was twofold. First, it was an early run, meaning that most of its salmon returned prior to the opening of commercial fishing. In addition, the river's salmon followed a more direct migration route, which avoided a large percentage of commercial netting berths to the east in zones 6, 7, and 8.

Evaluation of this information reinforced the strategy, noted above, of interpreting scientific results in a way consistent with one position versus another. Only this time, the interpretations varied between groups of commercial fishers as well. Ignoring the special explanation offered for the Conne River, commercial fishers in zones 3, 7, 8, 10, and 11 used the Conne River assessment to reinforce their view that there was no stock decline. A fisher just east of the Conne River put it this way: "See, even their own scientists says there isn't any problem. We been fishing here for 150 or more years, and salmon are more plentiful some years than others, but there ain't no crisis except in the minds of anglers." Fishers from zone 3 in the northwest part of the island heard of the Conne assessment through union meetings and used it to defend their fishery as well. In the words of one:

> They says we're catching too many fish and that we're hurting the rivers in Labrador and Quebec. I don't see it. I admits we must be catching fish from somewhere else, 'cause we ain't got many rivers 'round here. But hell, they've got just as big a fishery on the south coast, and they ain't even hurting their own rivers. I just don't see how we can be doing all that harm.

In contrast, fishers from the other zones generally saw the assessments as confirming much of their own recent experience. As the following quotation from a Gander Bay fisher suggests, the data simply gave credence to their own conclusions:

> Yes sir, I'm glad they [biologists] finally did the work. I knows the anglers have been yelling for a few years now, and to tell you the truth, anybody fishing for a living would have seen that there's fewer fish — particularly bigger ones. I thinks they should shut 'er all down for several years.

Fisher Participation in Union Activities

The second external factor relevant to understanding the divergence in fishers' assessment of salmon populations is the extent of participation in union activities. Newfoundland and Labrador fishers are represented by two unions, the Newfoundland

Fishermen, Food, and Allied Workers (NFFAW) and the Union of Fishers and Commercial Workers (UFCW). The former is by far the most important, representing in excess of 90 percent of the province's inshore, midshore, and trawler fishers, as well as most processing plant workers. The NFFAW has led the defense of commercial salmon fishers in confrontations with other user groups, particularly recreational associations. The position of the NFFAW has been that there is no widespread crisis in salmon populations (though individual rivers have been acknowledged as suffering) and that current proposals cloaked in conservation rhetoric are really attempts to reallocate the resource.

Although membership in the NFFAW is widely distributed throughout all salmon management zones, active participants on committees pressing the union's view on salmon come disproportionately from zones 3, 7, 8, 10, and 11. For example, the primary Canadian government consultative body for salmon management is the Atlantic Salmon Advisory Board (ASAB). Meeting once or twice annually to review fishery results, it reviews government management and offers advice on change. Typically, there are five to eight salmon fishers as well as two full-time union employees sitting on the board at any one time. Membership is replaced every three years. In the last nine years, only three out of twenty-six Newfoundland and Labrador commercial salmon representatives resided in zones other than 3, 7, 8, 10, and 11.[7]

A major consequence of union participation has been adherence to the union's position. When a fisher not only participates in the union but also represents other fishers in consultations with government at a table containing other competitors for the fish, the pressure to echo the union position is particularly strong, as indicated in this comment from an ASAB member from zone 3: "Sure, there may be the odd river with some problems, but this whole fight is about a way of life and basic human rights and the dignity of fishermen. Those _____ recreational guys just want everything for themselves, even if they destroy us."

It is important to note that the three ASAB members from zones 13, 4, and 5, while publicly supporting the union's position, acknowledged significant declines in their own catches and felt a bit more uncomfortable in the polarized debate. A fisher from zone 5 expressed his thoughts this way:

> Yes sir, it's a lot more work to catch some salmon. They's certainly less now, but maybe it's sort of a natural cycle. You know salmon are kinda funny. Some years there's lots and others there's not. I also think there are some anglers who want most of the fish for themselves. There ought to be some way we can deal with salmon declines without getting into these shouting matches and all this politics.

In this case, the fisher's direct experience at least left open the possibility that salmon stocks in his area were in decline.

It is impossible to determine how representatives to the ASAB were selected. Interviews with government officials suggest they were nominated by the union. Interviews with union officials suggest they were more or less randomly picked by government representatives. Regardless, union participation in general and membership on the ASAB in particular have been important in reinforcing the view that the resource is not endangered. Given the extensive informal communication linkages that exist between fishers in a region, the high level of union participation by salmon fishers from zones 3, 7, 8, 10, and 11 has meant that the union's position has been almost unanimously endorsed by others within those areas.

From Indigenous Knowledge to Perceptions of Stock Health: An Overview

The link between indigenous understanding of a resource and the fisher's assessment of that resource's health has been shown to

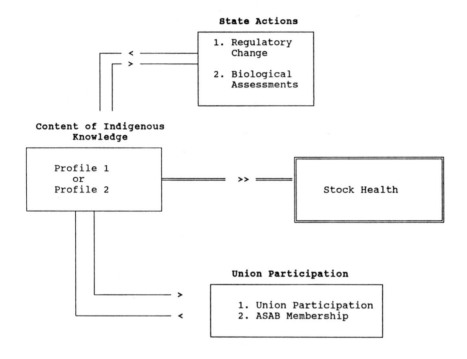

Figure 10.3. From indigenous knowledge to stock assessment.

be a process more complex than mere extrapolation from everyday experience. This process is summarized in Figure 10.3.

Fishers denying any significant stock problems begin with a highly instrumental, fragmented understanding of the resource for the reasons outlined in Profile 1. This understanding makes any overall judgement of stock condition difficult. Combine these conditions with government regulatory changes that have severely reduced any capacity to judge decline, and fishers are left with smaller harvests and considerable frustration and cynicism.

Finally, extensive participation with a fishers' union that is battling other users, primarily recreational groups, over allocation reinforces the denial of stock difficulties. As suggested by the bidirectional arrows in Figure 10.3, for those fishers perceiving the issue as a political conflict, there is considerable feedback in support of their perceptions coming from the unions, with each element reinforcing the other.

For fishers supporting the position of stock decline, the content of the understanding of the resource is much different. In addition, the impact of state intervention and union influence is much less relevant. Fishing more discrete, local stocks, these fishers have acquired a more highly integrated understanding of the resource. Additionally, because they harvest a much larger percentage of a smaller population of fish, any changes in abundance are more immediately recognized. With few exceptions, their personal experiences suggest that salmon populations have significantly declined. State regulatory changes have had little, if any, effect on their fishery, and the biological assessments indicating that salmon stocks have declined are consistent with these fishers' experiences. Finally, their much lower level of participation in union activities removes considerable pressure to assume a position contrary to their experiences. The result of this is that, while gauging their talk carefully so as not to repudiate or disavow union positions publicly, their support for the union's interpretation is at best tepid. Their concern, as indicated by the following quotation, lies in rebuilding stocks without having to give up large amounts of fish to recreational users:

> Well, you know the union got to protect us. I guess that's why they're saying there's really no problem. I wishes they'd stop denying and accusing. Hell, ask anyone around here, and they'll tell you salmon is some scarce these past few years. We need to stop poachers, seals, and everything else that's doing them in. I'm afraid that if they close it down completely, they'll [the government] never let us back at it. . . . But there's no doubt

if we don't do something, there ain't gone be nary a fish for anybody.

Indigenous Knowledge and Marine Management

The incorporation of indigenous knowledge into some form of cooperative management will be far more subtle and complex than one might first consider. Having said this, it still seems clear that the advantages will outweigh any disadvantages. In this concluding section, a few suggestions derived from the analysis of the Atlantic salmon fishery are offered for the integration of indigenous knowledge into fisheries management.

The first consideration relates to the great diversity, even contradictions, found between (and within) groups of fishers. It has been argued that even fishers pursuing the same species, sharing similar cultural patterns, living within a generally similar region, and utilizing identical gear can have enormously varied understandings. Although this chapter has focused upon important differences between two profiles of indigenous knowledge among salmon fishers, a careful examination would reveal subtle, and potentially important, differences within each profile as well. Sensitivity to the issue of variation even within homogeneous groups is a necessary starting point if fisher knowledge is to be incorporated into management policy.

If localized knowledge is not monolithic, whose knowledge, if anyone's, do managers use? In order to answer this question, it is critical to fully understand the social processes through which the knowledge is produced. In current discourse, we must be able to deconstruct it, to understand how fishers come to believe the things they do. Of particular importance are processes linking diverse, and often fragmented, indigenous knowledge to conclusions about a resource. For management purposes, the most critical conclusions are likely to be (1) how many of a species exist at a certain point in time and (2) whether such numbers will allow sustainable harvesting at current levels of exploitation. As is clear from this analysis of salmon fishers, knowledge of the process turns out to be as important as the content of the knowledge itself in understanding fishers' conclusions about the resource. Without dismissing the importance of the knowledge-producing process for other fisheries, a reasonable generalization might very well be that the more a fishery is characterized by user-group conflict, cash payments for catches, and significant state regulation, the more critical the knowledge process itself becomes.

Following from this assertion, it is important for researchers and fisheries managers to differentiate detailed knowledge about the resource from any conclusions drawn about that resource. This is particularly true when conclusions may be potentially colored by heavy politicization. Even where conclusions appear inconsistent with several other sources of information, including scientific assessments, and are therefore likely to suffer in their credibility, much of the knowledge itself may be valid and useful. An interesting case in point relates to fishers in zones 10 and 11. For several years the governments of France and Canada have argued over ownership of the North Atlantic off the south coast of Newfoundland. By virtue of owning the islands of St. Pierre and Miquelon, France has long claimed a 200-mile proprietary zone. Canada has countered with a much more restrictive twelve-mile zone. In submitting its case for international arbitration, the Canadian government was particularly sensitive to various stocks resident and migratory to the contested area. Lacking scientific knowledge about the major migration paths of Atlantic salmon through the disputed area, several experienced Canadian salmon fishers from the area were asked for information on salmon migration. The information was allegedly incorporated into the Canadian submission to an international court, which would hopefully minimize future salmon interception by French nationals. Fisher information was used despite the fact that the fishers in question had long criticized Canadian fisheries scientists for claiming that there were significant declines in salmon populations. Detailed drawings of migration paths provided by these fishers fit very closely with limited tagging experiments undertaken by fisheries biologists.

A final concern suggested by many of the interviews, which has not been developed in this analysis, might be termed *knowledge devaluation* resulting from contact with state biologists. Although the culture of inshore salmon fishers is far less submissive and deferential than it used to be toward outside "experts," there is still a noticeable tendency for fishers to defer to pronouncements made by highly educated government scientists. When confronted with professional pronouncements that run counter to their own experiences, many fishers simply shake their heads, walk away, and perhaps utter a few profanities under their breath. One noticeable change that is worthy of future research consideration is the tendency among fishers having regular interaction with state managers and scientists to adopt the latter's discourse. This tendency is particularly noticeable for fishers who serve on government advisory committees. Consider the following conversation, recorded at a local fishers' committee, in which a local salmon fisher serving on a government committee attempts to explain to another fisher — who is also his cousin — why quotas needed to be introduced in 1990 for the first time:

> I knows b'y. But you have to understand the situation. The government has twelve rivers around the island they calls index rivers. On average over the last five years, spawning escapement has allowed only 40 percent of the necessary egg deposition needed. As well, the last two years have had unusually cold seawater temperatures. This has meant a much higher smolt mortality as well. The bottom line is that catches need to be reduced, and they thinks a quota is the best way to do it. You may not like it, but you can't argue with their information, so what you gonna do?

Note the references to index rivers, spawning escapement, egg deposition, and smolt mortality. This is the language of the research biologist. According to interviews, the only fishers to employ such language are those with considerable experience on government advisory committees. There appears to be little doubt that serving on such committees distances these people from their neighbors. In the words of one fisher attending the above meeting, "He's been too long upalong with 'em. He even talks like one."

The specific consequences of such assimilation are not entirely clear. It may do nothing more than remove such converts from positions of influence among fellow fishers. If widespread enough, it is possible that the generation and legitimacy of indigenous knowledge may decline as more and more fishers become assimilated. If fishers come to think, act, and understand in such an alien discourse as a result of participation in resource management, it could make the integration of indigenous knowledge into such management more difficult. Certainly this appears to be an area worthy of further investigation.

The primary objective of this examination of indigenous knowledge among commercial Atlantic salmon fishers has not been to dismiss the existence and importance of indigenous knowledge for sustainable management. The concerns raised in the analysis suggest that the integration of fishers and their knowledge into more truly cooperative management schemes will not be as straightforward as initially believed. This in no way suggests that the general objective is flawed. Fishers possess a rich and detailed understanding of those species they hunt, an understanding including the larger marine environment in which their quarry dwells. There are numerous ways in which such knowledge is relevant for sustainable management practices. Perhaps the overall message of this analysis is that enumerating and codifying such knowledge is not enough. Before it can be fully utilized for sustainable management, the social conditions and constraints under which it is produced must also be appreciated. Then, hopefully, it will be possible to utilize fishers and their knowledge to protect and manage the marine environment, and, through such efforts, to sustain fishers themselves and their communities.

Lessons for Modern Fisheries Management

1. Sensitivity to variation in knowledge is a necessary starting point for incorporating indigenous knowledge into fishery management.
2. The more that fishers exploit local, nonmigratory stocks, the more useful indigenous knowledge will be for stock management.
3. The greater the extent to which a fishery is characterized by migratory stocks, user-group conflict, market transactions, and extensive state regulation, the less likely it is that local fishers' knowledge will be monolithic.
4. Where localized knowledge is divergent or contradictory, it is critical to fully understand the social processes through which knowledge is produced. In other words, how do fishers come to believe what they do?
5. It is important for researchers and fisheries managers to differentiate detailed indigenous knowledge about the resource from any conclusions about its health.
6. Knowledge devaluation — the process whereby fishers come to discount their own experiences and invoke the discourse of state managers — may be a significant result of their ongoing participation on state-sponsored advisory committees.

Notes

1. Social scientists utilize the terms *indigenous* and *folk knowledge* to describe the accumulated meanings, symbols, and understandings fishers utilize in their harvesting activities. For a worthwhile discussion of some possible distinctions, see McCorkle (1989) and Thrupp (1989).

2. Biological assessments carried out by Canadian government biologists concur with the assessment indicating serious stock decline. As a result, on March 6, 1992, a five-year moratorium was placed on the commercial salmon fishery on the island portion of the province of Newfoundland and Labrador and a compensation package was offered to any salmon fishers willing to permanently relinquish their license.

3. These numbers are derived from interviews with Canadian Department of Fisheries and Oceans scientists and from statistics provided by this ministry. To estimate numbers of fish commercially harvested, annual landings recorded in metric tons (1 mt = 2,200 lbs.) were noted and converted into numbers of fish by dividing the tonnage expressed as pounds by the average size of fish landed (approximately five pounds).

4. The relationship between the catching capability (efficiency) of gill nets and the population of the resource to be harvested appears to be curvilinear rather than linear where the total harvest represents a small percentage of the population. As a result, harvests do not rise and fall in any systematic way with population changes. Depending upon how dense the migratory schools are (a function of water temperature, food supply, and certain other variables in addition to total population size), a given level of harvesting effort may catch approximately the same number of fish even though the total population of fish migrating through may have declined substantially.

5. Tagging experiments conducted in the late 1960s and early 1970s suggest that between 5 and 15 percent were of distant origin. Ice conditions and seawater temperature appear to be major determinants of timing and pattern of migration. This contrasts with 85 to nearly 100 percent in zones 3, 7, 8, 10, and 11. Although more recent tagging experiments have not been undertaken, there is no evidence that the general pattern is any different today. (Note: the preceding data is from personal communications with scientists at the Canadian Department of Fisheries and Oceans.)

6. These respective commercial sector exploitation rates of 65 percent and 80 to 90 percent were provided by the Ministry of Fisheries and Oceans, Canada. They are based on smolt and adult tagging studies carried out on a number of provincial rivers.

7. The author sat as a member of the ASAB from 1982 until 1992 and was able to keep an ongoing tabulation of the residences and management zones of board members.

References

Berkes, Fikret
    1977   "Fishery Resources Use in a Subarctic Indian Community." *Human Ecology* 5:289–307.
    1987   "Common-Property Resource Management and Cree Indian Fisheries in Subarctic Canada," in Bonnie McCay and James Acheson, eds. *The Question of the Commons.* University of Arizona Press: Tucson, pp. 66–91.

Cordell, John
1974 "The Lunar Tide Fishing Cycle in Northeastern Brazil." *Ethnology* 13:379–392.

Dasmann, R. F.
1974 "Ecosystems." Paper presented at the Symposium on the Future of Traditional Primitive Societies, Cambridge, England.

Department of Fisheries and Oceans (Canada)
1991 Background paper for 1991 Atlantic Salmon Advisory Board. Science Branch, DFO: Ottawa.
1992 Canadian Atlantic Fisheries Scientific Advisory Committee. Advisory Document 92/3. Science Branch, DFO: Ottawa.

Faris, James
1972 *Cat Harbour.* Institute of Social and Economic Research: St. John's, Newfoundland.

Firestone, Melvin
1967 *Brothers and Rivals: Patrilocality in Savage Cove.* ISER: St. John's, Newfoundland.

Klee, G. A., ed.
1980 *World Systems of Traditional Resource Management.* John Wiley and Sons: New York.

Kloppenberg, Jack
1991 "Social Theory and the De/Reconstruction of Agricultural Science: Local Knowledge for an Alternative Agriculture." *Rural Sociology* 56(4):519–548.

McCay, Bonnie
1981 "Optimal Foragers or Political Actors? Ecological Analyses of a New Jersey Fishery." *American Ethnologist* 8:356–382.

McCay, Bonnie, and James Acheson
1987 *The Question of the Commons: The Culture and Ecology of Communal Resources.* University of Arizona Press: Tucson.

McCorkle, Constance M.
1989 "Toward a Knowledge of Local Knowledge and Its Importance for Agricultural Research, Development, and Education." *Agriculture and Human Values* 6(3):4–12.

McGoodwin, James
1990 *Crisis in the World's Fisheries: People, Problems, and Policies.* Stanford University Press: Stanford, CA.

Morrill, Warren T.
1967 Ethnoicthyology of the Cha-Cha. *Ethnology* 6:405–416.

Nakashima, Douglas J.
1990 "Application of Native Knowledge in EIA: Inuit, Eiders, and Hudson Bay Oil." Canadian Environmental Assessment Research Council: Ottawa.

Palmer, Craig
(Forth-coming) "Cod Traps, Fish Plants, and Changing Attitudes Towards Women's Property Rights," in P. Sinclair and L. Felt, eds. *Living on the Edge.* ISER: St. John's, Newfoundland.

Pinch, Trevor
1986 Confronting Nature: *The Sociology of Solar-Neutrino Detection.* D. Reidel: Dordrecht, Netherlands.

Pinfold, Thomas
1990 "An Economic Statement on Developing the Atlantic Salmon Resource in Newfoundland and Labrador." Gradner-Pinfold Associates: Halifax, Nova Scotia.

Porter, Marilyn B.
1983 "Women and Old Boats: The Sexual Division of Labour in a Newfoundland Outport," in E. Garmanikow, ed. *Public and Private: Gender and Society.* Heinemann and BSA, London.
1985 "She Was Skipper of the Shore Crew: Notes on the History of the Sexual Division of Labour in Newfoundland." *Labour* 15:105–123.

Sider, G.
1989 *Culture and Class in Anthropology and History.* Cambridge University Press: Cambridge, England.

Thrupp, Lori Ann
1989 "Legitimizing Local Knowledge: From Displacement to Empowerment for Third World People." *Agriculture and Human Values* 6(3):13–24.

11

Regulating Fjord Fisheries: Folk Management or Interest Group Politics?

Svein Jentoft and Knut H. Mikalsen

Fisheries management is a highly controversial issue in most coastal states, and Norway is no exception. The uncertainty of scientific data is but one problem (Bailey and Jentoft 1990). Another is the lack of consensus on what constitutes critical and relevant knowledge (Smith 1990) and whether "folk knowledge" should play a role in regulatory decision making. Still another problem lies in the fact that management decisions in most cases have distributional effects that do not satisfy the Pareto optimum criteria: that is, if there are winners, there are also losers. Consequently, the proficiency of management regimes is not simply a question of precise knowledge — whether scientific or folk — but also one of equity and justice.

A rational approach to fisheries management would seek whatever information is relevant and available. However, in democratic societies, regulatory decisions are vulnerable to political pressure, lobbying, manipulation, logrolling, and even sabotage. In the fisheries, as in other sectors of the economy, this is an integral and "normal" part of the regulatory process in most countries.

Thanks are due to Gunnar Trulsen and Ernst Bolle, both with the Regional Fisheries Administration in the county of Troms, for access to their archives and for allowing us to attend the 1991 meeting of the Regulatory Committee. Thanks also to Geir Andreassen of the Fishermen's Union for information and support, and to Hans-Kristian Hernes and Jacob Meløe for comments on an earlier version of this chapter.

Therefore, better knowledge will not necessarily change the course of events. "Knowledge is power," the saying goes, but not always.

One institutional response to this dilemma, a response advocated by an increasing number of social scientists, is the co-management model (Jentoft 1989, McCay and Acheson 1987, McGoodwin 1990, Pinkerton 1989). *Co-management* means "a shift away from autocratic and paternalistic modes of management to modes that rely on the joint efforts of traditional fisheries specialists and fishing peoples" (McGoodwin 1990:189–190). Ideally, co-management gives user groups a real influence, in the sense that their practical knowledge makes a difference in the decision-making process. At its best, co-management leads to consensual management. The ideal implications of this are summarized in Figure 11.1.

The benefits of co-management are expected to be (a) greater participation by user groups in decision making, which enriches the regulatory process by providing a broader base of information and knowledge, (b) an increased rationality in the regulatory process as participation produces more adequate and legitimate regulations, (c) the positive effects that participatory democracy has on legitimacy as it alleviates the centralistic bias of fisheries management, and (d) with increased legitimacy, an enhanced adherence to rules and regulations by user groups contributing to a more proficient management regime.

There are, however, problems and pitfalls in this co-management model. At its worst, co-management may lead to co-optive management, that is, to the co-optation of user groups and local interests rather than to efficient participation on their part. Another problem is the definition of user groups (Pàlsson 1991). Should user groups be defined exclusively in functional terms, or should their definition also be extended to include, for instance, any local groups affected by regulatory decisions? Thus, co-management may enhance folk management, but not necessarily the participation of all affected interests. In such a case, the information pool may then be too narrow, and the data that are defined as relevant

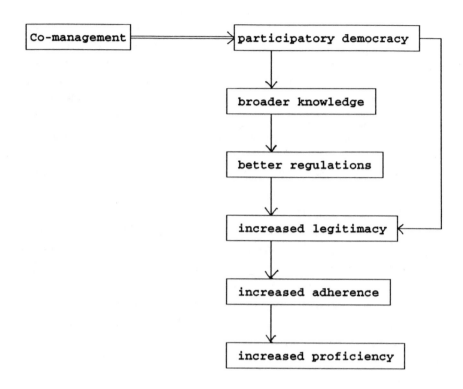

Fig. 11.1. The ideal implications of co-management.

may be biased to the advantage of some user groups and at the expense of others.[1] Therefore, the effectiveness of co-management may depend upon the degree of inclusiveness in the formal institutions established for the purpose of co-management.

In a previous paper, we have described the system of user-group participation in fisheries management on the national level in Norway (Hoel, Jentoft, and Mikalsen 1991). Norway, however, also has a tradition of co-management at the local level. Jentoft and Kristoffersen (1989), for instance, have described the history of fishers' participation in the regulation of the Lofoten fishery. In this chapter, we describe a more comprehensive system of local fisheries management — one that has been established to regulate the fisheries in the many fjords along the Norwegian coast. This system has increasingly been attacked as undemocratic and ineffective, and there have been suggestions that it needs to be restructured. "The billy goat should not guard the oatmeal sack" is the standard argument here. The fishers should not have exclusive control over the content and scope of local regulations, as they have now.

There is evidence to support this argument, and we shall return to it later. The main problem with this co-management system, as we see it, is the privileged position of functional groups at the expense of territorial, or local, interests. The circumstances and consequences of this are the central theme of this chapter.

We start with a brief description of the development of the system, with particular attention to organizational form and decision-making procedures. We then describe the system in operation, reporting from the 1991 meeting of the Regulatory Committee in the county of Troms (northern Norway) and focusing in particular on the way in which decisions are made and justified. Then we discuss the problems and dilemmas of fjord fisheries management, with special emphasis on the ability of the present system to tackle new challenges raised by political claims from municipal authorities and ethnic groups and by new forms of marine production such as sea ranching.

The Roots of the Present System

Important things to keep in mind when discussing local regulations in Norwegian fisheries are the rugged coastline, with its many branching fjords (see Figure 11.2), and the long-standing tradition of combining small-scale fishing and agriculture among people living along the fjords. Historically, the fjord fisheries for cod have been the exclusive domain of the local population, which uses traditional gear such as hand lines, longlines, and gill nets. In

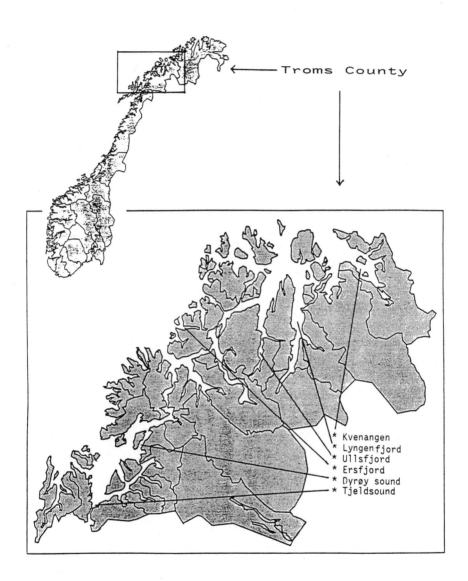

Fig. 11.2. Fjords of Troms County.

northern Norway, fjord resources have been of considerable importance to a particular ethnic group, the coastal Saami, who have
regarded these as their property in common and who have asserted
that these resources should be managed on the basis of their
detailed knowledge of the fjord as an ecological system (Bjørklund
1991, Eythorsson 1991, Paine 1965).

The introduction and diffusion of new harvesting technology
such as trawls, purse seines, and the power block were increasingly
seen as a threat to traditional adaptations along the fjords. From
the mid-1950s there was a steady stream of local demands for
protection against "intruders," coming mainly in the form of demanding restrictions on trawling and purse seining in some of the
fjords. Three kinds of restrictions were on the agenda:

1. Closing all, or parts, of the, fjords to certain types of gear.
2. Dividing the fjords among fishers or boats using different gear.
3. Allocating time slots within different areas to various user
   groups.

These demands came from local groups and individuals, including fishers, farmers, and local branches of the Fishermen's
Union. In the county of Troms (see Figure 11.2), demands for
restrictions prompted the regional branch of the Fishermen's Union
to appoint a task force in 1954 to look into the matter. In its
subsequent report the task force listed and evaluated some fifteen
such demands and recommended that all should be met.

The director of fisheries[2] and the Ministry of Fisheries, however, did not approve — for reasons of control as well as "justice"
— and the demands were shelved in anticipation of a general
review of questions pertaining to the handling of similar cases. The
demands, from Troms as well as from other parts of the country,
did much to speed up the search for appropriate decision-making
procedures. The question of procedures was then taken up in
discussions between the Ministry of Fisheries and the Fishermen's
Union, and agreement on a regulatory system was reached by the

end of 1958. From these discussions, a formal standing organization was created, which included representatives from the Fishermen's Union and the Ministry of Fisheries. This marks the beginning of the Advisory Committee on Local Regulations, hereafter called the Regulatory Committee. The formal procedures adopted were roughly as follows:

1. Demands for local regulations, from groups or individuals, were to be referred to the regional branch of the Fishermen's Union (through its local association) by June 1 each year.
2. An advisory committee including representatives from all relevant gear groups was to be set up in each county along the coast and chaired by the regional inspector of fisheries.[3] The director of fisheries could choose to be represented by an observer.
3. All demands were to be evaluated and decided upon by the committee, and additional information would be collected whenever necessary. The committee's decisions were to be heard by the regional branch of the Fishermen's Union and then eventually submitted to the director of fisheries.
4. Final approval for local regulations was to be a matter for the Ministry. Cases of wider significance or questions of principle were to be referred to the Fishermen's Union before a decision was made.

In addition to formal procedures, two substantial criteria, or premises, for handling the regulatory demands were laid down: first, the approval and implementation of local regulations were to be subject to the rules of the general fisheries legislation as defined in the so-called Salt Water Fishing Act of 1955. This act delineates the forms of regulations that can be enforced. Local restrictions on access were thus not to be considered a special case.

Second, open access was to be the general rule. Restrictions for particular groups would only be enforced when considered necessary to protect or enhance the resource or to secure "rational and responsible" harvesting within the different fisheries. The fjord resources were still to be counted as common property rather than the exclusive assets of the local population.[4]

Representation and Legitimacy: Calls for Change

There were no changes in these procedures until the late 1970s. The director of fisheries did, however, at some point propose a reduction in the number of representatives on the committee — his motive obviously being to cut costs.[5] Nothing came of this, however, and the committee continued to operate with its original structure until 1978. To be sure, the Ministry of Fisheries did put the system of local regulations on the agenda when reorganizing the Fisheries Extension Service in 1973, but there were strong pressures from below to retain the established procedures. The regional inspector of fisheries in Troms stated that he considered it crucial that the final decisions on local regulations are based on advice and proposals from institutions where the various gear groups are represented and where local knowledge is prevalent.

There were also pressures for change from other quarters. Of particular interest was a proposal from the Troms branch of Norway's biggest political party, the Labor Party, advocating stronger representation for the local communities affected by the committee's decisions. In a statement from a fisheries conference held in 1975, the party pointed to the growing conflict between local fishers and "intruders" over fishing grounds and proposed what it called a democratization of the regulatory system. This was to be achieved by bringing the municipal authorities into the decision-making process. The gist of the argument was that such a change would force a better balance between the needs of local communities and the interests of functional groups when regulations were decided.

As rumours began to spread — through the press and other channels — that the Ministry was considering changes along these lines, the fishers' organizations passed resolutions opposing any major revisions of the system. One local association stated that given the procedures currently being worked out, whereby the municipal authorities will have a strong influence, the fishermen have every reason to fear that other considerations will weigh more than the real interests of the fisheries. Others followed suit, and the fishers' uncertainty propelled the Ministry to put the question of revisions of the traditional system on the agenda. A proposal for new decision-making procedures was put forth in 1976 and was enacted during February 1978. The changes were minor, the most important being that the County Council, a popularly elected assembly, was granted the right to appoint a representative to the Regulatory Committee.

The new procedures did nothing, however, to solve the problem of local considerations versus functional interests. In response, in 1987, the Ministry proposed changes in the way demands for local regulations would be handled. The purpose was to secure better representation for the local communities and the fjord fishers. One of the more interesting proposals here was the suggestion from the chief executive of the Regional Fisheries Administration in Troms that the committee should be discontinued and that its functions should be taken over by the politically appointed County Fisheries Board.[6] The outcome, however, was that municipal authorities were given the right to propose local regulations, which heretofore had been more or less the exclusive domain of the local branch of the Fishermen's Union. In practice this gives the municipality the right to act on behalf of individuals or groups within the local community. In addition, the number of representatives per gear group was cut from two to one. The formal structure of the current system is shown in Figure 11.3.

There are now three courses of action open to those who want to demand new regulations of the fjords: (1) they can make the local branch of the Fishermen's Union take up the demand and carry it forward to its regional branch, where it will be considered by the board before being sent over to the Regional Fisheries Administration, which, in due course, summarizes the statements and puts the issue before the Regulatory Committee; (2) they can forward

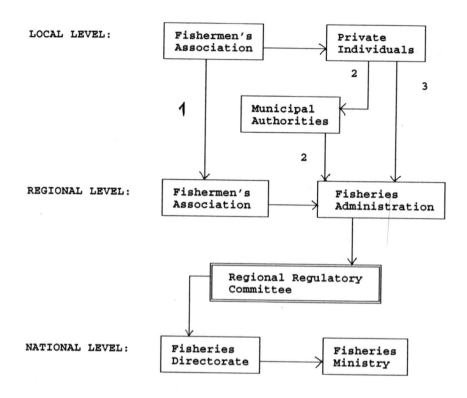

Fig. 11.3. Current system of co-management for fisheries in Troms.

demands through the municipal authority, which, usually with support and recommendation, takes it to the Regional Fisheries Administration and then eventually to the Regulatory Committee; or (3) they can raise their demands directly with the Regional Fisheries Administration, which then seeks the views of the regional branch of the Fishermen's Union before preparing the agenda for the Regulatory Committee.[7] Whatever route is chosen, the decisions of the Regulatory Committee are submitted to the director of fisheries, who then takes them to the Ministry for final approval.

Another issue in the debate on decision-making procedures has been the representation of ethnic groups with long-standing traditions in fjord fisheries, such as the Saami population (Bjørklund

1991, Paine 1965). Saami organizations have made several demands during the 1980s for representation and influence in regulatory decision making. A proposal for closing all the fjords of Finnmark and Troms to active gear has met with little enthusiasm from government and the Fishermen's Union. The question of Saami representation on the advisory committee has been on the agenda but has yet to make an impact on the formal structure of the system.

The latest call for significant changes has come from several municipalities, some of which have a considerable Saami population. More or less identical resolutions pertaining to the composition of the advisory committee have been passed, all calling for broader and more democratic representation in regulatory decision making.

The pressures for change are mainly justified by pointing to the restrictive policy of the committee in enforcing local demands for regulations of the fjords. There is, on the face of it, much to be said for such a viewpoint, as indicated in Table 11.1.

Table 11.1 The outcome of local demands in Troms County, 1959–1991

	demands	granted	rejected	% rejected
1959–1969	100	45	55	55
1970–1979	124	38	86	69
1980–1981	84	36	48	57
Total	308	119	189	61

Table 11.1 shows that nearly two-thirds of the demands for local regulations have been rejected by the committee, a fact that, taken at face value, certainly indicates a restrictive policy.[8] One should, however, note that the system is extremely liberal in the sense that the same demands can be made again and again, which quite often happens, without any solid documentation as to the need for new restrictions. So the official interpretation is that the system is rather generous, because local demands have been met

without the necessary documentation being provided. There is, furthermore, an understanding among the members of the Regulatory Committee that further restrictions would be tantamount to closing the fjords to all but local fishers.

This viewpoint, however, is strongly disputed on the local level, where the legitimacy of the present system is clearly wearing thin. The problem, as perceived from "below," is the lack of attention being paid to the problems of the fjord fishers and the particular needs of the local communities.[9] There is certainly a grain of truth in this, as we observed firsthand when we attended the 1991 meeting of the advisory committee.

Co-Management in Practice

The Advisory Committee on Local Regulations usually meets once a year, and its 1991 meeting was held on December 20. At that meeting the committee had seven members, including the chief executive of the Regional Fisheries Administration, who chairs the meetings and prepares the agenda. There was also one fisher delegate from each gear group (e.g., hand line, longline, trawl, gill net, and Danish seine), appointed by the regional branch of the Fishermen's Union. One additional member had been appointed by the County Council, but it was not quite clear whom he was supposed to represent. This individual happened to be a hand-line fisher and was the only member who actually lived in a fjord district. The other members came from various communities along the coast of Troms. In addition to the chief executive, two of his deputies were present. Also attending as an observer at this particular meeting was an officer from the Fisheries Control Service. The deputies and the observer were active in the debate but did not vote. Also present at this meeting was the director of the Fisheries Research Institute, who had been invited to brief the committee on one particular issue.

too-detailed regulations could prove impossible to enforce and control. The observer from the Fisheries Control Service then exclaimed, "I get goosebumps from listening to you guys. There is no way we can control the depth of the seine."

The meeting was now nearing a vote. The chief executive summed up the debate and proposed a vote on two alternatives: (1) banning all active gear, with the exception of the saithe seine, within the existing limits, that is, as far out in the fjord as proposed by the research director and (2) banning all active gear including the saithe seine, as well as small-mesh nets, within the same area. When the vote was finally taken, the second option was supported only by the county representative.

There are several things worth noting regarding the nature of this debate. First, it was a real discourse, where arguments were raised and challenged on the basis of functional knowledge of the particular fisheries in question. The pros and cons of the request were seriously considered, and the vote eventually resulted in a different solution than the one proposed by the Fisheries Research Institute and backed by the municipal government, the local fishermen's associations, and the chief executive. It is also interesting that the delegates of passive gear (hand line, longline, and gill net) voted in favor of the seine. In other words, the classic conflict in Norwegian fisheries between active and passive gears was superseded. A third observation is the minimal attention that was paid to the demands of resident Ullsfjord fishers. It was briefly mentioned that there were also local seiners to consider, but the demands and criticism expressed by the local fishermen's associations did not seem to inform the debate. Notably, these associations did not want an exception for the saithe seine.

The next issue on the agenda was a demand for the prohibition of the saithe seine on the inner part of the Kvenangen Fjord (see Figure 11.2), which had been raised by the local branch of the fishermen's association. It was argued that larger vessels from other districts operating on the fjord were catching not only pollack but also large quantities of cod. Municipal authorities supported their demand and argued that the local net fishery for pollack was being damaged by the intruding seiners. In a later remark by the Regional Fisheries Administration, these arguments were disputed because the pollack, unlike the fjord cod, is not a stationary resource.

In the following debate, the county delegate supported the local claim. One of the deputies contended that this case boiled down to a question of competition, and his principal, the chief executive, agreed that this was a classic conflict of interest. "This fishery is the only alternative for some of the smaller seiners," the trawler delegate added, indicating that the local fishers were not the only legitimate interest in this case. This ended the discussion. When the vote was taken, the county representative was the only one in favor of the local demands.

The third demand related to the Ersfjord (see Figure 11.2) and had been raised several times before. A local fishers' association wanted to ban the use of shrimp trawls on the fjord, claiming that in reality the shrimpers were there to catch cod. It also wanted the seines banned because of alleged pollution problems caused by seiners dumping herring. In his comment following the case for the demand, the chief executive noted that there was no proof of such dumping, that it is illegal in any case, and that it should be dealt with by the appropriate authorities. Neither was there any evidence that seiners were destroying cod fry. He also doubted that the trawlers were there for reasons other than harvesting shrimp.

The debate followed the same pattern as before: the county delegate supported the local demand; the other delegates refuted it. As the county delegate stated, "We lack a thorough scrutiny of the matter, as always." The chief executive agreed that the documentation left much to be desired. The seiner representative was asked whether he knew the area. It had been a long time since he had fished there, so there was not much he could say. The longline delegate, contending that the argument for the local demand was insufficient, said, "If we shut one fjord because of illegal dumping, we will have to shut them all."

The fourth issue was similar to the previous ones and concerned the Dyrøy Sound (see Figure 11.2). It had been raised by a private individual with the support of municipal authorities. Again, the county delegate made the opening statement, asserting that "active gear should be kept out of such areas!" The chief executive found that the evidence cited to justify the demand again left much to be desired, and he feared that similar demands for other sounds might be forthcoming. In response, the county delegate exclaimed that "local demands are never given a fair hearing by this committee!" The observer from the Fisheries Control Service then added, "In this case, we are talking of large seiners, and they can do a great deal of damage." These statements did not, however, impress the other delegates, and there was again a majority in favor of rejecting the demands.

The fifth issue concerned the Tjeldsound (see Figure 11.2), and the debate more or less repeated itself. Almost routinely, the county delegate made the opening statement and again supported the local demand. The trawler representative was next to take the floor and said he found the issue difficult due to the scarcity of information about local conditions. He suggested, however, that the seiners were usually both careful and considerate. "That I doubt," retorted the county delegate. Next, the chief executive said he again found the evidence in support of the demand wanting, to which the county delegate replied: "And then we go against what local people say they have observed. We think that we know, but in reality we, too, have interests to defend." As before, only the county delegate voted for prohibition, and the demand was rejected by the usual majority.

The last issue on the agenda was an "old-timer" and concerned the closing of the Lyngenfjord (see Figure 11.2) to all types of active gear. The issue had repeatedly been raised by three adjacent municipalities and now, as always, was strongly supported by the local fishers' associations, political parties, and Saami organizations.

Leading off, one of the deputies made it clear that there were also shrimpers and Danish seiners from a fourth adjacent municipality fishing on the fjord. In addition, there was a local seiner operating in the same area. The trawler delegate said that he used to fish on the Lyngenfjord and that he knew it well. He went on to say that he could accept a ban on Danish seiners because it would affect very few fishers. He was, however, afraid that such a decision would lead to similar demands from other fjords and communities. "Exactly!" retorted the county delegate, implying that the trawler fisher was really speaking for himself. After this, the committee decided to reject the local demand but agreed to ban the Danish seine from the fjord. This time the vote was unanimous, and the county delegate justified his vote by referring to a new grid that had proven effective in separating cod from shrimp.

In summation, there was an obvious lack of scientific data on the state and development of the fjord resources that might have informed the decision-making process. There was also some evidence to support the claim that the Regulatory Committee tended to ignore local information and demands, and observations by local people usually were either disputed or disregarded. Moreover, available information on local claims and conditions was seldom referred to in the discussion, and there seemed to be a certain confusion as to whose responsibility it was to provide and present it: was it the responsibility of those who initiated the demands or rather a task for the Regional Fisheries Administration, or should it fall to the Regulatory Committee itself? Knowledge of local conditions and adaptations was thus erratic and secondhand, brought in by chance on the basis of the experience of any member of the committee who happened to have been fishing in the area. None of the present delegates were residents of the fjords in question. Indeed, with the exception of the county delegate, they all came from coastal communities.

The popular notion that this was an instance of "the billy goat guarding the oatmeal sack" was also shared by some members of the committee. They were there, they felt, mainly to protect their own interests, that is, the interests of the group they represented. The restrictive policy of the committee, however, might also have been due to the attitude of the Regional Fisheries Administration, for whom the overriding principle is open access. The statement by the administration that the regulation of fjord fisheries was now more or less complete indicated a negative attitude toward further restrictions.

The principle of open access, however, can hardly be regarded as neutral, as its adoption has tended to benefit "intruders" with active gear such as seines or trawls. Because scientific data was largely missing, any claim that stocks were in fact local and in jeopardy due to overfishing could thus be disputed in the committee.

A serious problem with the regulatory process, of which the Regulatory Committee is an important part, is the uncertainty as to what the goals of the system really are, or what they should be. Are the objectives of local regulations purely ecological, that is, to protect the stocks from overfishing? Or are they also socioeconomic in the sense that the regulations should be enforced with a concern for how they will affect the social and economic viability of local communities? As the county delegate said, "It is important to have something to hold on to, a social and political principle. Too much is left to our own discretion. In short, there is too much 'bingo' going on here."

Local Regulations: Rational Management or Partisan Politics?

It is, as indicated by our description in the previous paragraphs, difficult to muster official support for local demands for the closing of fjords to mobile gear. The Regulatory Committee has clearly adopted a restrictive policy, and there is little, if any, interference from other regulatory agencies in matters of fjord management. Furthermore, there are no formal procedures of appeal other than the restating of demands from one year to the next. The restrictive practice of the committee does not, however, include existing regulations, as these, if necessary, are extended more or less automatically. In this sense there is a certain conservative bias in the system, one that seems to reinforce the restrictive attitude among the members of the committee. The moral seems to be that one should be careful in enforcing new restrictions because once they are in place, they are virtually impossible to abolish.

The tendency to opt for the status quo also suggests that there are obvious conflicts of interest involved, such as those between the largely stationary fjord fishers and the more mobile seiners and trawlers. It also reflects the prevailing view that the fjord resources are the common property of all fishers, rather than the property in common of the local population only. The basic policy has been one of no discrimination on grounds of tradition, residence, or ethnicity. This has had consequences for the definition of user groups as well as for conceiving what constitutes relevant and valid knowledge for regulatory purposes.

Basically, there are three types of knowledge involved in fisheries management at the local level: first, scientific knowledge (research-based information on the state of the local marine ecosystem); second, functional knowledge (expertise concerning the efficient operation of particular types of gear and technology and knowledge concerning when and where harvests are taken in the local fjords); and third, territorial knowledge (practical knowledge of the fjord resources as well as of the significance of particular adaptations for the economic viability of local groups and communities).[11]

Biological information has hitherto played a minor role in the regulation of the fjords, even though there is an obvious need for more research into the state of "local" stocks. At virtually every meeting of the advisory committee, there have been complaints about the lack of scientific data and calls for research. In this sense it is probably wrong to conceive of local regulations as resource management in the strict sense of the term. This is because the knowledge base is currently far from firm enough to facilitate any form of proper management.

What we have labelled as functional expertise is much more prevalent, as it is virtually the main basis of representation on the committee. This type of know-how is closely related to economic interests and is thus less neutral than, for example, biological expertise. Thus, although functional knowledge is, in itself, objective, it can easily become biased when it is used to promote the economic interests of particular gear groups.

To the extent that regulatory decisions are based chiefly on the advice and views of the members of the committee, the present system is as much an arena of interest-group, or partisan, politics as a tool for the rational management of the fjord resources.[12] There is a fundamental problem of democracy and equity here, because, as we have seen, the demands for regulations come mainly from individuals and groups who are not paid-up members of the Fishermen's Union (and who are therefore without genuine representation in the committee), and decisions are thus made without the input of a group that may have vital interests at stake. In this sense the fjord fishers simply are not considered a user group, which goes a long way in explaining their lack of influence within the system.

Territorial knowledge and local interests are, in other words, poorly represented in the regulatory process. As it is, knowledge of the specific fjords in question will tend to be accidental and superficial. Those representing the functional groups may, of course, have knowledge of the territory in question, but this knowledge originates from shorter visits to the fjords.[13]

Traditionally, information about local conditions used to be obtained through so-called field trips by the committee, that is, through visits by committee members to the fjords and communities in question. However, these trips became less frequent over the years and no longer take place. One could, of course, presume that delegates of traditional gear groups (hand line, longline, and gill net) would identify with the fjord fishers and thus provide the latter with some form of representation. These delegates, however, are not necessarily well-informed about the particular problems of the small-scale fjord fisheries, and they also, as we have seen, tend to go along with their more "modern" counterparts on the committee.

Information about local problems and adaptations is brought to the committee's attention mainly by the representative appointed by the County Council. The individual occupying this position during our study obviously defines his role as a defender of local and ethnic interests. His "mandate," however, is unclear, and it does not oblige him to act in this manner. Furthermore, he has not been able to make any real impact on policy, despite being well-informed and articulate, even after considerable experience with grass-roots politics.

Inadequate representation, therefore, is one aspect of the problem because it excludes important user groups, restricts the flow of information within the system, and leaves the regulation of the fjords largely to the more "professional" segments of the industry. On the other hand, the organization of regulatory policy making does reflect the more profound and prevailing conception of local resources as common property, that is, as resources that are open to any Norwegian citizen. On this point conventional wisdom and formal legislation collides with the local conception of these resources as the property primarily of the local community, property that is held in common and obtained through centuries of use. This local perception of ownership or priviliged access applies to fish stocks that reside in the fjord as well as to fish that migrate into the fjord.

The overriding problem then, is how to strike a balance between, on the one hand, the goal of protecting local adaptations and customary rights and the need to observe the principles of formal legislation and, on the other, the legitimate interests of functional groups.

Conclusion

The management of the fjord fisheries in Norway satisfies many of the criteria for co-management status: it is a bottom-up rather than a top-down process; the approach is participatory in the sense that user groups play an active role in decision making; and government authorities are involved at various stages of the regulatory process, but otherwise they keep a low profile. Nevertheless, the system has come under increasing pressure from various user groups, especially from fishers residing in the fjords, and from other quarters such as local authorities and Saami organizations. This demand for reform results largely from a general and persistent neglect of local demands. In other words, what started as a dissatisfaction with the substance of the regulations has by now turned into a negative sentiment toward the regulatory process itself. The main problem, therefore, at least as it is defined by its critics, is the selection of representatives and the choice of decision-making procedures. We have seen the same tendency on the national level where pressures for expanding the basis for representation have led to changes in the composition of the National Regulatory Council (Hoel, Jentoft, and Mikalsen 1991). Similar demands might well result in the inclusion of new groups at the regional level.

This controversy over representation pertains to some of the basic questions and dilemmas that are inherent in functioning democracies. For instance, who are the affected interests and how should they be represented in the democratic process? Or in other words, "Who ought to be a member of the demos?" (Dahl 1989:4). In our case, membership in the demos seems to be founded on functional criteria and lamentably excludes local groups and individuals, among whom the harvesting of local stocks has been, and still is, an integral part of their livelihoods. There is, however, a drive to include interests other than fisher groups in the management of the fjord fisheries, such as in the case for management at the national level. What is now being suggested is a more exocratic form of regulation (Tivey 1978), entailing a stronger participation by groups and institutions from outside the industry. The afore-mentioned proposal to abandon the Regulatory Committee alto-gether in favor of the politically appointed County Fisheries Board, is a case in point. There are, however, limits to representation in the sense that the inclusion of new participants for reasons of legiti-macy may reduce the decision-making efficiency of the system.

Another classic problem in democracies pertains to scale. What is the natural locus of democracy: the local, the regional, or the national level? And what are the problems and benefits of choosing one level over another? Managing fisheries in a context of regional diversity, technological competition, and political con-flict raises the question of power and jurisdiction: At what level should regulatory decisions be taken and enforced? Is the current regionally based system the most appropriate one for managing the fjord resources? Or could one instead envisage further decentrali-zation through the delegation of regulatory power to local groups? Notably, this question has not yet been raised.

On the other hand, one may also envision a more active role for central government agencies in the regulatory process. Local conflicts over space and harvesting rights raise the question of arbitration, which could mean stronger involvement on the part of central government. As it is, further decentralization seems a more likely option. As the reader will recall, the entry of municipal authorities was proposed but rejected in the mid-1970s. The pro-posal has, however, been raised again by three municipalities in the county of Troms, and this time they may have a better case. For one thing, marine biologists now confirm what fjord fishers have argued for years: that there is, in fact, such a thing as local stocks. The cod stocks in the fjords and in the Barents Sea (spawn-ing in the Lofoten district) are different species, but the present system of fisheries management is based exclusively on stock estimations pertaining to the Barents Sea cod. As a basis for the regulation of the fjord fisheries, this system has in effect been shown to be obsolete, and from now on, fjord fisheries management will require better knowledge of local conditions. This can be achieved by reintroducing the old practice of field trips, engaging the regional and local branches of the Fisheries Extension Service, or by direct representation of local user groups and municipal authorities in regulatory decision making.

Secondly, in some areas sea ranching may become commercially viable. A government task force appointed to analyze future trends in Norwegian aquaculture foresees a dramatic increase in the use of this new technology, suggesting that the fjords will serve as excellent sites for this industry.[14] Until now attention has been focused largely on the biological and technological obstacles while neglecting the possible legal and organizational problems of sea ranching. There is, however, a growing awareness of the regulatory problems that may arise surrounding such questions as who will cover the costs of cultivation, who will be allowed to harvest, and how will the scope and forms of participation be decided? And if there is going to be open access, the problem of "intrusion" could prove a lot more difficult to solve than it is today. There is no reason to believe that local authorities and user groups will allow the current state of affairs to continue in a situation where the fjord resources are becoming both more plentiful and more accessible.

A third issue concerns the rights of the Saami. Norway has ratified international legislation to protect and advance the interests of aboriginal peoples (for instance the International Labour Organization convention no. 169). Norway has also agreed to the interpretation that preserving the culture of such groups must, if necessary, include the protection of the material foundation upon which the culture rests. In 1990 the newly elected Saami Parliament (Sametinget) raised the question of the implications of this legislation for fisheries management. The Norwegian government promptly responded by asking a distinguished law professor (since appointed head of the Norwegian Supreme Court) to give an expert opinion on the matter. He concluded that a more direct participation of Saami user groups in regulatory decision making is in order and, furthermore, that discrimination beneficial to Saami fishers and their communities is legally defensible if necessary to sustain Saami culture. The Ministry of Fisheries accepted this, and beginning in 1992 the Saami Parliament was authorized to appoint one delegate to the National Regulatory Council. It follows logically, one may argue, that similar representation should be provided at lower levels.[15]

These challenges call for adjustments of the fjord regulatory system. The principle of open access will be difficult to maintain, and the management task will become much more complex. One virtue of open access is that it is simple administratively and politically. Rules and regulations can be minimized, standardized, and made obligatory for all participants on equal terms. If, however, each fjord is to be treated as a unique ecosystem, and fisheries management is to help sustain the economic and social viability of local communities, the present management system is ripe for reform. Thus, the current system of functional representation and knowledge must be modified through the broadening of participation along territorial and ethnic lines.

However, although a broader representation may result in a more enlightened debate in the committee, it will not necessarily change the course of events. New knowledge will not automatically have a decisive effect on the decisions made, as the members of the committee are representatives of group interests rather than local fjord communities. This suggests that even if local knowledge were brought to the committee, it would not make much difference as long as there was no recognition of the first right to the fjord by those who live there.

A rational debate on the organization of regulatory decision making will also presuppose a clearer definition of objectives, of what is to be achieved through local regulations. Even if, as is presently the case, regulations are conceived as exceptions to the general rule of open and equal access and are enforced on grounds of resource conservation, there is obviously still a case for consulting the functional groups affected. This is so because restrictions would infringe on their rights by limiting the use of particular kinds of harvesting technology. If, on the other hand, open access is rejected and the principle of first right is accepted, and regulations are seen primarily as a legitimate means of protecting the traditional adaptations and economic viability of fjord communities, there is a much stronger case for including representatives of local communities and, in some cases, localized ethnic groups. The choice is ultimately a political one, and before it is made, further debates about the substance and forms of regulatory decision making seem futile.

Lessons for Modern Fisheries Management

1. Ethnic and territorial groups in fisheries should be identified and their unique cultural and political histories considered in management decision making.
2. Fishing communities should be treated as unique ecosystems, with a fundamental management goal being to sustain their economic and social viability.
3. The consultative integration of local folk knowledge in management decisions can increase the legitimacy of those decisions.

Notes

1. See Jentoft 1989 for an extensive treatment of the pros and cons of co-management.
2. The director is the head of the Directorate of Fisheries — essentially a professional, or "staff," institution that provides expertise and advice to the Ministry. It is beyond doubt a highly influential institution in that capacity.
3. The inspector was the head of the Regional Fisheries Administration and was ultimately responsible to the Ministry of Fisheries in Oslo. Today the committee is chaired by the chief executive of the regional branch of the Fisheries Extension Service, a position broadly comparable to the fisheries inspector of the 1950s.
4. It is interesting to note that the fisheries inspector, at almost every meeting of the committee, took great care to remind the members of this principle of

open access. The consequences for the committee's decisions are difficult to assess but do indicate a restrictive interpretation of its mandate.

5. The committee did a lot of on-the-spot investigations during its first years to provide more accurate information about local conditions and to hear the groups involved. These field trips tended to deplete the committee's budget, and the director was anxious to put a stop to this.

6. The County Fisheries Board consists of five members appointed by the County Council (Fylkestinget). According to guidelines set by central government, the board shall "deal with all questions of relevance to the fisheries within the county, and supervise the Fisheries Extension Service and other government initiatives for the promotion of the fisheries." As for fisheries regulation, the board can be consulted by central government, and its advice has in some cases been decisive. One example is the distribution of licences for aquaculture and small-scale dragging.

7. If the first or third route is chosen, bypassing the municipal authorities, the municipal authorities are always consulted by the Regional Fisheries Administration before the case is submitted to the Regulatory Committee.

8. We find the same tendency for other counties. Eythorsson (1991), for instance, finds that all demands for local regulations in the county of Finnmark for the years 1986, 1988, and 1990 were rejected, with one exception.

9. In everyday speech the distinction between the two categories — fjord and coastal fisher — tends to be blurred. Here the term *coastal* refers to the outer coast, including the islands and their shores but excluding, and in contrast with, the fjord. We therefore distinguish between fjord fishing (which is not simply fishing in a fjord, but fishing in a fjord by a fisher who lives along a fjord) and coastal fishing.

10. There are no quota regulations specific to the fjords. There are, however, national quotas that also apply to the fjord fisheries. These quotas are decided solely on the basis of estimations of ocean stocks. In the county of Troms, the following local regulations are in place: a ban on shrimp-trawling, which applies to thirty local areas; prohibitions on shrimp-trawling during the night in a few other areas; a ban on the use of the purse seine, Danish seine, and other seines in six local areas; and a conditional ban on these gears in nineteen other areas.

11. There is no clear-cut distinction between the last two categories. The efficient operation of, say, a shrimp trawler cannot take place without some prior knowledge of the territory, that is, of the when and where of trawling. Consequently, where there is functional knowledge such as how to handle a shrimp trawl, there is also territorial knowledge. As our colleague Professor Jacob Meløe has pointed out in a personal note to the authors, "since you will have to locate the shrimp before you can try to catch it, you will also, in

learning shrimp trawling, learn a great deal about the when and where of shrimp harvesting." He also argues that the reverse is true: that functional expertise is a prerequisite for acquiring territorial knowledge. Knowledge about the different stocks of fish in the fjord, and about where and when they are to be found, has been acquired through people's fishing there for generations. The reason that you cannot have one type of knowledge without the other, he argues, is that fishing gears are artifacts, man-made objects. With artifacts you can always ask, What do we use that for? If there were no ways of using the artifact, it would not exist. In other words, fishing gear is used to catch fish, and in order to catch fish you need knowledge of gear, as well as territory.

12. The concept of "partisan politics" refers to a situation where a decisionmaker "makes decisions calculated to serve his own goals, not goals presumably shared by all other decisionmakers with whom he is interdependent" (Lindblom 1965:29).

13. Fishers other than those living along the fjord will actually be "foreigners," "nomads" roaming from fjord to fjord. The fjord fisher is, on the other hand, a settler with a far richer and more profound knowledge of, for example, spawning grounds and seasons and of the carrying capacity of local stocks. The "foreigners" come only to extract, and they acquire only the knowledge that is needed for that purpose.

14. Det nasjonale utvalg for havbruksforskning: Perspektivskisse for norsk havbruk. Norges Fiskeriforskningsråd 1990.

15. There is clearly room for cooperation and alliances between municipal authorities and Saami organizations when it comes to articulating demands for local regulations, as illustrated in the statement by one mayor that "we have achieved something because we have cooperated with the Saami organizations."

References

Bailey, Conner, and Svein Jentoft
   1990   "Hard Choices in Fisheries Development." Marine Policy (July):333–344.

Bjørklund, Ivar
   1991   "Property in Common, Common Property, or Private Property: Norwegian Fishery Management in a Saami Coastal Area." North Atlantic Studies 3(1):41–45.

Dahl, Robert A.
   1989   Democracy and Its Critics. New Haven, CT: Yale University Press.

Eythorsson, Einar
    1991   "Ressurser, livsform og lokal kunnskap." University of Tromsø, Institute
            of Social Science, master's thesis.

Hoel, Alf H., Svein Jentoft, and Knut H. Mikalsen
    1991   Problems of User-Group Participation in Norwegian Fisheries Manage-
            ment. Paper presented at the Second Annual Meeting of the Interna-
            tional Association for the Study of Common Property. Sept. 26–29.
            Winnipeg: University of Manitoba, Natural Resources Institute.

Jentoft, Svein
    1989   "Fisheries Co-Management: Delegating Government Responsibility to
            Fishermen's Organizations." Marine Policy (April):137–154.

Jentoft, Svein, and Trond I. Kristoffersen
    1989   "Fishermen's Co-Management: The Case of the Lofoten Fishery." Hu-
            man Organization 48(4):355–365.

Lindblom, Charles E.
    1965   The Intelligence of Democracy: Decision Making Through Mutual Ad-
            justment. New York: The Free Press.

McCay, Bonnie J., and James M. Acheson (eds.)
    1987   The Question of the Commons: The Culture and Ecology of Communal
            Resources. Tucson: University of Arizona Press.

McGoodwin, James R.
    1990   Crisis in the World's Fisheries: People, Problems, and Policies. Stanford,
            CA: Stanford University Press.

Paine, Robert
    1965   Coast Lapp Society, Vol. 2. Oslo: Universitetsforlaget.

Pàlsson, Gìsli
    1991   "Ændrede forutsetninger i islandsk fiskeripolitik," in J. Williams (ed.):
            Fiskerireguleringer: rapport fra Nordisk Kontaktutvalg for fiskerispørs-
            mål. Seminarrapport 516. Copenhagen: Nordisk Ministerråd.

Pinkerton, Evelyn (ed.)
    1989   Co-Operative Management of Local Fisheries: New Directions for Im-
            proving Management and Community Development. Vancouver: Uni-
            versity of British Columbia Press.

Smith, M. Estellie
    1990   "Chaos in Fisheries Management." Maritime Anthropological Studies
            3(2):1–13.

Tivey, Leonard
    1978   The Politics of the Firm: Government and Administration. Oxford:
            Martin Robertson.

12

Summary and Conclusions
Evelyn W. Pinkerton

Introduction

A decade ago co-management was an exciting new concept full of promise and possibilities — part of a new paradigm in fisheries management. Most scholars and practitioners of co-management felt they were battering against the doors of old models and institutions and receiving little recognition or understanding.

Today, in some circles, the documentation of co-management and folk management may have outstripped the development of a theory that places these paradigms in perspective. The theory exists within different disciplines but has not been connected systematically enough to the description of these community-based regimes. The paradigm has therefore landed in the embarrassing position of receiving sufficient recognition to be considered — at least in some quarters — more the norm than the exception. That is, when folk management regimes are not present, not working, have been overwhelmed, or show no evidence of appearing, some scholars now feel that this must be explained or even that the new paradigm is thereby challenged.

This state of affairs means that it is time to revisit some of the basics. This discussion therefore focuses on a few key considerations that need to be more thoroughly incorporated into co-management theory. The first consideration is, What are the fundamental minimum conditions under which one could expect folk management or co-management to arise? and the second, What are the main vulnerabilities of a co-management or folk management system to being undermined?

Basic Conditions for the Development of Folk Management

We should not assume that the development of folk management is automatic and inevitable; rather, we should assume that a long period of stable population size, location, and resource use is required as an opportunity for local populations to experiment, learn, and adapt to local environments. The study of cultural evolution suggests that folk management systems were probably developed through trial and error in situations where the resource was fairly forgiving, that is, mistakes could be made that were nonfatal to species. Alternatively, in very fragile ecosystems such as reefs surrounding atolls, it was possible to observe the impacts of overharvesting and to adjust activities in a timely enough fashion. Local populations tended to develop myths, rituals, taboos, and prohibitions regulating use; at a minimum they developed customary usage to prevent the recurrence of mistakes. Gradually, as cultural ecologists might put it, local groups developed resource use patterns that kept them within the long-term carrying capacity of the local land and marine areas.

There are fishing communities that have not been and are not now in a position to develop folk management practices, for any of several reasons. In some cases, their historical situation and the time of development of the fishery have not put them in a position to learn the carrying capacity of the stocks. For example, transplanted frontier communities may be established at a time of great resource abundance relative to local population. Industrial development or integration into world markets may happen too rapidly for these populations to learn resource limits and how to harvest in keeping with them. In addition, regional or national population explosions may occur too rapidly for local groups to learn how to exclude outsiders effectively.

McGoodwin's discussion of the development and loss of *respeto* in Chapter 2 is an excellent case in point. Local populations had developed a system of space allocation that worked under certain technological and population conditions (nonmotorized vessels, low population density). Space allocation had proved effective as an indirect conservation tool in communities with relatively stable populations (e.g., Martin 1979). But this system had only a few decades of existence before motors and a flood of newcomers overwhelmed the fishery. This situation was quite different from the one in Buen Hombre (Chapter 5), where a long-accumulated body of local knowledge, a long-established set of practices contributing to conservation, and a conscious conservation ideology gave the local community a basis from which to develop resistance to the impacts of new technology and the invasion of outsiders. Furthermore, the impacts were neither as sudden nor as great as those in Mexico.

In Newfoundland, also in a "frontier" community, as documented by Palmer in Chapter 9, the timing and rate of resource decline relative to local understanding of the dynamics of the decline were fortuitous. Lobster fishers were beginning to understand that their fishing practices caused the decline at the same time that state conservation measures were brought in. State measures were therefore welcomed as a solution to a problem the fishers recognized. How perfect a solution these measures will be in the long run is unclear, but at least for the present, they seem to answer a serious problem. It is possible that Newfoundland lobster fishers will come up with observations in the future about new trends in abundance patterns and suggest new directions for management. The current coincidence of similar perceptions will likely prove to be more the exception than the rule, unless exceptional leadership is exercised in communication between the state and the community.

Leadership from the community depends on the willingness, usually related to ideological preadaptation, of local fishers to take on tough problems. Taylor (1987) documented a situation in which

Irish salmon fishers were faced with resource decline but were still unwilling to participate in management activities such as local enforcement. Through the fishers' experience with British colonialism and distant Irish landlords, the local tradition of hatred of remote authorities that claimed ownership of the local resource was more powerful than traditions of self-management. Because of this, they were unable to organize themselves to accept mutual monitoring through an authority of their own choosing and eventually lost the resource. These situations remind us that the local ability to develop and implement a management regime depends a great deal on this readiness and the local cultural understandings and ideology that support it.

In other words, folk management either may not develop in the first place, or it may develop and be overwhelmed. If some scholars are beginning to assume folk management is the norm, it is time for the theory to spell out more clearly the conditions under which one does not find folk management, or aspects thereof. This book contributes new material to this discussion. In the next section, we begin considering the second question: the points of vulnerability at which established folk systems may be undermined.

The Co-Management Spectrum: Accommodations to State and Market

Let us assume for the moment, as this book's editors argue, that folk knowledge and folk management methods, where they exist, make a valuable contribution to achieving sustained yield and equity in fisheries management today when they are a part of the management system (see Introduction). In some cases, they constitute a viable management system on their own, that is, a system of self-management. In other cases, when folk knowledge and local perspectives are incorporated into a larger management system as co-management, they may make the difference between the system's having legitimacy or not, having local relevance or not, and in general operating more rather than less effectively.

Let us also assume that senior governmental regulatory agencies are not going to disappear and that folk management systems will survive in most cases by working out an accommodation with senior levels of government. This accommodation can take place at many points on a spectrum, as discussed below.

Let us further assume that most folk management systems today operate in the context of world markets, which can have a powerful impact on the operation of the folk system. Markets external to the local world of folk management have the potential to either threaten and undermine the local system or to enhance it, if local institutions can interact with the market on terms favorable to them.

The key theoretical problem asks: Under what conditions does an accommodation of folk systems to the state and the market take place without destroying the benefits of folk management? What is the optimum role for the state, and the optimum relationship with the market for the preservation of benefits? One way to illustrate the structural dynamics that underlie all accommodations is to pay attention to negative accommodations that illustrate the points of vulnerability of co-management arrangements. In these instances, some or all of the potential benefits are destroyed through (a) incorporation of the folk system on terms that strip it of political power so that it is outvoted or overpowered by regional or state interests in competition with it for local stocks, interests that have less stake than the local community in conserving those stocks; and (b) co-optation by the "free" market, by special competing interests, or by state bureaucracy itself. That is, the folk system is integrated in such a way that individuals are incorporated as components of the larger market system while shedding obligations to be accountable to sound resource management and equity within the folk community. This may introduce or reinforce structural inequality and also lead to elimination of the resource.

Such negative accommodation can also lead to absorption of the folk system by the state system and to the loss of benefits.

The theory needs to spell out more clearly the conditions under which such negative accommodations can happen, so that assumptions are not made that folk management or co-management systems are automatically beneficial accommodations. Some examples in this book illustrate negative accommodations and provide hints about some of the structural reasons for them.

The Range of Local Accommodations to the State

We can now represent more fully the range of possible accommodations in several ways: for example, along a scale of more or less involvement by the state. This section sketches the characteristics of a few points along the scale, their vulnerabilities and strengths (including the incorporation of folk management in the unfavorable terms described above). What are some of the problems and dilemmas encountered at various points in the range?

The range of types and possibilities for local/state accommodation can be expressed in at least two ways, depending on what aspect of the situation is being illustrated. Different models are appropriate to different situations and problems.

One way is through a simple horizontal continuum from nearly total self-management to nearly total state management, with various forms of co-management in the middle (see Figure 12.1). This type of schematic emphasizes the balance of initiative, influence, and power exercised by state and local forces. Who controls key management resources such as the generation of data about the resource, the moral authority, or the legitimacy? Who makes the critical decisions about the resource? There is no assumption in this model about where power fundamentally lies. There are several kinds of power in any regime; some kinds of power may rest entirely with a local regime that generates the

Self-management ─────────────────────────▶ State management

1 2 3 4 5 6 7 8 9 10

(Points 2 through 9 are forms of co-management, with the state or the community having different degrees of control or involvement.)

Fig. 12.1 The horizontal range of local/state accommodations.

information and the moral authority needed to manage, as well as the ability to initiate problem solving.

Stoffle, Halmo, Stoffle, and Burpee (Chapter 5) document a situation that began at point 1 on this scale and demonstrated the chief vulnerability of that position. Buen Hombre fisheries are managed by a local association that to a large extent has limited access to the resource by outsiders, indirectly controls production by the fishers, controls habitat-threatening activities, and perpetuates a conservation ideology that reinforces the other controls. The local community acts as a corporate group whose members have primary responsibility to the collective interest of the group. Because of its isolation, this local system has been effective for a long time but recently has lacked the power to defend itself from increased pressure from outsiders who fish using illegal technology in Buen Hombre waters. Although the local regime was legitimate in the eyes of its members, locals obviously feared that the use of violence to protect local waters might not be condoned by government.

Eventually a limited form of co-management evolved when government recognized and supported the right of Buen Hombre fishers to exclude outsiders from their waters. The participation of government touches no area of management except the provision of badges to local enforcers and the willingness of the Coast Guard to hold offenders arrested by Buen Hombre patrolmen. This places the Buen Hombre management system at approximately point 2 on the scale, because this support is largely moral and symbolic.

This situation exemplifies a class of community-based regimes that are in co-management situations merely for protection from outsiders of the self-managing mechanisms they already have, not because they need other resources from government. Their "accommodation" with the state is simply to invite protection in order not to be invaded, so that the regime can continue in a sustainable and biodiverse manner.

Equally characteristic of this class of regimes is their ability to access data produced by outside, university-based researchers to support their claims. This use of external resources to legitimize their activities and assert their moral authority before governments (e.g., Freeman 1989, McCay 1989) is key to their success. One might be tempted to consider this in itself a very limited form of co-management of another management activity: data collection and analysis. However, the data was not needed for local management so much as for legitimizing the defense of the area. Unless this data and outside legitimation also help the local association to censure the occasional insider offender who uses destructive technology (a distinct possibility), it should probably not be considered a co-management adjunct.

Another way of conceptualizing the range of local/state accommodations is to view the government as holding all of the legal authority, moral authority, and actual power but voluntarily allowing aspects of management to devolve to local bodies according to very discrete and specific agreements on certain management functions. This is the top-down, vertical "contracting out" model of state management and co-management (see Figure 12.2). The basic model most closely resembles Arnstein's "ladder of public participation," adapted by Berkes, Preston, and George (1991) to apply to co-management, to describe a range of degrees of participation. It can be usefully modified to represent a tree of different functions or activities that can devolve all the way to community control of this function or to any midpoint. This model views power as moving from the state to communities.

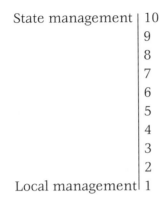

Fig. 12.2 The vertical range of local/state accommodations.

On this scale the Dalmatian oyster fishers in Louisiana described by Dyer and Leard (Chapter 3) would be at point 2. Their system of self-management is completely enclosed in a cultural enclave that regulates access to the fishery, production levels, enhancement (reseeding), and distribution of products. The system has been sustainable and stable over long periods, and it has been protected by long-term leases from the state that effectively create oyster "fiefs": the oyster fishers have territorial rights over most of the suitable habitat for oyster production. From the perspective of this model, the local system operates effectively because the state has provided legal arrangements that enable it to operate.

In reality, the system is more interactive than this model implies. The Dalmatians arrived in Louisiana in the mid-nineteenth century and helped develop the state leasing law and the Oyster Commission fifty years later. It might therefore be more accurate to say that the preexistence of the self-management system led to the creation of laws to support and protect it. The main difference between this situation and the one at Buen Hombre is that the Dalmatians' relationship to the state is far more formalized and long-standing. The state of Louisiana has legally delegated management authority to the leaseholders for most management functions. Their system operates independently according to its own principles inside the state system.

State Management and Control of Power

In either the horizontal or the vertical model of power sharing, it is important and relevant not to automatically view the state as a neutral bureaucracy that operates rationally in making agreements for the devolution of management activities or for the sharing of power. Government resource-management agencies may instead be organizations that respond to bureaucratic rules that have less to do with positive management outcomes than with self-maintenance and control. And they are likely to reflect the different training and different cultural construction of management problems that Ward and Weeks document (Chapter 4). In the Texan oystering industry, establishing relations with fishers via the Oyster Advisory Committee is viewed by the state as serving conflict-resolution goals, not as a way of incoporating local knowledge into management. The proprietary attitude of the state toward the definition of management problems and the control of data leads it to hold its own data secret, to assert its own definition of what causes oyster depletion and how this is properly measured, and to exclude the fishers from the process as much as possible. This management system could be rated at approximately 9 on the vertical scale.

It is probable that management authority was delegated to the Dalmatian Louisiana oystering community not because of any desire to include local knowledge, but rather from a desire by the state to incorporate a conflict-free system that was stable and highly productive and made few demands on state resources. Long-term leases would be classified ideologically as quasi-private property, and the management relationship more resembles a private-property system than it does a delegation of authority. Dalmatian oyster fishers and Buen Hombre fisher communities resemble each other in their degree of self-management and independence from state resources. However, they differ greatly in the degree of power each holds in its relationship to the state. This influence is explored in the next section.

Client Group Influences on State Management and Co-Management

Whether government agencies share much or little information or power with fishing communities, client politics can play a key role and directly or indirectly impact the development or operation of co-management arrangements. That is, agencies are influenced by the needs and demands of the groups they "serve," especially those that derive the greatest economic benefits from the resource and are the most politically powerful (Lowi 1969, McFarland 1987). Client groups influence agencies by lobbying governments to make the agency do what they want or by directly lobbying the agency. In many cases, the clients are considered the most knowledgeable political constituents about resource needs, as well as industry needs, and are likely to get at least some of what they want. In extreme cases, they capture the agency, which is then less sensitive to the desires and needs of other groups or to discovering what the general public interest might be (e.g., the wisest and most sustainable use).

If agencies are captured by major economic or political groupings, especially those with a statewide or nationwide influence, they are unlikely to develop effective co-management relationships with fishing communities (Pinkerton 1992). The Dalmatian community in Louisiana is an unusual case of a folk community that became the major client group. This was able to happen because the community became established early in the history of Louisiana, had preexisting knowledge and experience (when there may have been few political competitors), and was therefore able to set much of the political agenda regarding oyster management.

The leasing law and commission they promoted has allowed them virtual control both of the best habitat and of policy making.

The demands of the more powerful client groups often drive the information needs of agencies and thus help define what data agencies are likely to consider most relevant when making decisions. Felt (Chapter 10) reminds us that ecosystem-based communities — which might otherwise be in a position to develop co-management relations with agencies — will have difficulty being heard in the political shadow of an established client group that currently drives policy at the regional or national level. Scholars have stressed the importance of the ecosystem-based local territory or watershed as the basis for local knowledge and self-management (Blomquist 1992; Feit 1988; Berkes 1989; Berkes, Preston, and George 1991). However, it has not been sufficiently emphasized that such groups are likely to be at a disadvantage in the political arena and may possibly be in direct competition with other groups for the same fish. Felt's analysis of the intra-union dynamics — in which ecosystem-based fishers tend to follow the union line, even when they know better — adds an important new twist to the competition between differently situated fishers and adds ideological loyalty to the explanation for the nondevelopment of co-management at the local level.

The impediment to co-management in this case is not simply the existence of more powerful clients. It is also the manner in which the state constructs representation on advisory committees. Here Felt's case resembles the Norwegian example described by Jentoft and Mikalsen in Chapter 11, in which the statewide union is given a preeminent place on advisory committees and policy groups at several levels. The state has not allowed an appeal mechanism, and the interests of the ecosystem-based fjord fishers and the county are systematically outvoted by union representatives coming from a broader regional representation and having little or no knowledge of the fjord situation. Jentoft and Mikalsen question what the goals of the system really are. Apparently the goals are quite different for the differently situated fishers. The fjord-based fishers wish to conserve the local stocks. The "outside," or non–locally based, fishers wish to have greater access for all or higher yields for themselves and do not appear overly concerned about whether local fjord stocks could be over-fished. In this situation, the state appears to have chosen to exclude one group from policy making because of the established position of a more powerful client.

Both the Newfoundland and Norwegian examples provide important illustrations of impediments to the development of co-management that should also be seen as points of vulnerability of co-management arrangements to client politics. They suggest how pressures on state agencies could threaten existing co-management or self-management systems.

The Range of Local Accommodations to the Market

The prominence of these market and extralocal interest group influences in the modern world means that the local/state continuum really should include them, analyzing the circumstances under which they can be constructive or destructive to the aims and potential benefits of folk management.

In addition to client groups in competition for the resource or the raw product, the world market may affect the local/state accommodation either directly or through client groups at any point on the spectrum. At the self-management end of the spectrum, outside markets may directly and positively impact individuals so that they are motivated to create a self-managing system, as analysts of collective action have shown. Forming collective institutions is economically rational behavior, both from an individual and from a collective viewpoint, when scarce common pool resources are endangered by market competition. Alternatively, if local regimes cannot adequately control the behavior of their members, markets may fragment the behavior of local fishers. If fishers begin to perceive local resources only as commodities to be exploited in terms of discount rates and opportunity costs, rather than as parts of complex ecosystems that require sustainable and integrated management for the long-term benefit of the community or public, the local regime is in trouble. It will have to devote greater resources to disciplining these fishers, or the management system and the resource will eventually collapse. Of course, management systems work best when compliance is low-cost because of ideological or economic motivations. The following example illustrates how economic motivations can lead to low-cost compliance.

The southern California groundwater users' associations (Blomquist 1992, Ostrom 1990) formed in the 1940s were a response to the long-term costs of not acting collectively. Water users in particular areas faced saltwater intrusion into underground aquifers because of competitive and nonsustainable overuse of groundwater, high costs of litigation among competing users, and high costs of importing water if basins were overused. By reaching out-of-court agreements and jointly hiring watermasters to monitor and enforce the rules they agreed to, the many individual users achieved a guaranteed and sustainable supply of water at a much lower cost. They did this on a watershed-by-watershed basis, through public discussions among water users within a watershed. Key to these discussions were an examination of relative costs and a willingness to mutually monitor each other's water use through hiring an authority who would be accountable to all users, the watermaster. In this case, the high costs of importing water (external markets), in addition to the high costs of resolving conflicts over local water, drove watershed residents to invent a system of self-management. This is, of course, a positive and sustainable accommodation to the market, on terms that are favorable to local users and that are sustainable.

In contrast, Palmer's example (Chapter 9) of a Maine lobsterman who bends the trap limit rules to enlarge the family catch is an example of negative accommodation to the market. This lobsterman apparently is willing to view the immediate commodity benefits to himself as greater than the long-term benefits to himself, and to the community, of sustainable harvesting by the informal rules and practices. Either the lobsterman will be effectively censured by the community, or his actions will contribute to the erosion of the folk management system. Either way, his actions are a cost to the community. He is either raising the cost of collective account-ability, or he is depreciating the value (or institutional capital) of the folk management system.

Pomeroy's discussion (Chapter 1) suggests the possibility (al-though it is frustrated in her particular case) that local unions and cooperatives can positively mediate the relations between local users and external markets. National rules and national support for the clear jurisdiction and operation of cooperatives and unions can prevent the abuse of rules by local elites who might attempt to form advantageous relations with outside markets at the expense of other local users. This prevention of abuse appears to have been successfully executed by the local marketing co-op in Buen Hom-bre. Pomeroy's optimism that government authorities can facilitate and encourage the formation of such local institutions should be tempered by an analysis of the extent to which government is or is not captured by either competing user groups or external economic interests.

Summary of Hypotheses Generated

This chapter opened with two questions: (1) What are the fundamental conditions required for a folk management system to arise? and (2) What are the vulnerabilities of folk management and co-management systems to being undermined? It is now possible to generate a number of hypotheses from the foregoing discussion and from chapters in this volume. Rather than attempt to sort these into specific responses to one question or the other, it may be more useful to consider these hypotheses as general conditions that appear to be associated with successful regimes. A number of these conditions have been identified in some form or another in previous work (e.g., McCay and Acheson 1987; Pinkerton 1988, 1989, 1992; Berkes 1989; Berkes, Preston, and George 1991; Ostrom 1990). In those cases, they are confirmed, reframed, nuanced, or enriched here.

The benefits of folk management and co-management systems are likely to be optimized

1. where a local tradition has had the possibility of developing at all (e.g., not in Mexico as described by McGoodwin, or in Newfoundland as described by Palmer), whether undisturbed, as in Buen Hombre, or successfully transplanted, as into Louisiana from Yugoslavia (Dyer and Leard);
2. where local ideology, values, or recent experience creates a willingness to deal with collective problems such as hiring an authority or accepting the authority of the group through mutual monitoring (present in Buen Hombre and the California groundwater situation, absent in Taylor's Ireland, and apparently a problem in Palmer's Maine);
3. where local ideology and belief systems reinforce customary use practices, that is, where at least some practices are consciously conservation oriented (Anderson [Chapter 6], lack of it in Palmer's Maine);
4. where geographic boundaries are clearly defined (Pomeroy, all ecosystem-based cases by implication);
5. where the boundaries are based on an ecosystem in which the natural relationships can be easily observed and understood by fishing communities (Felt; Jentoft and Mikalsen; and Stoffle et al.);
6. where membership is clearly defined (Pomeroy);
7. where traditional community-based systems are relatively undisturbed and still function to insure conservation and equity (Stoffle et al., Dyer and Leard);
8. where the state supports locally based institutions and the development and unification of policy, or at least does not

impede them (Dyer and Leard; Stoffle et al.; not in Pomeroy), or where outside authorities legitimize a weakly supported local regime and thereby encourage state recognition (Stoffle et al.);

9. where state regulations are consistent with local understanding of resource problems or actual local practices (this is more likely to happen where communities have not devised their own systems of conservation that have proved effective in the past [Palmer's Newfoundland, not in Ward and Weeks]);

10. where the state is not subject to extreme pressure from non–ecosystem-based interest groups that are in competition for the same stocks (Jentoft and Mikalsen; Felt; Dyer and Leard);

11. where self-management by an encapsulated group serves the interests of the state in conflict resolution and in being relieved of management costs (Dyer and Leard; Stoffle et al.);

12. where local populations have the self-confidence to use confrontation and threats of violence to support their territorial claims (not in Stoffle et al. or Palmer);

13. where the state has not constructed representation only from nationally based organizations but has allowed locally based representation that is not bound by national organizational ideology to arise (Felt; Jentoft and Mikalsen);

14. either where there are individual economic incentives to cooperate because the costs of going it alone are higher, or where management is invested in knowledgeable local authorities with the power to make the rules and also implement them (Dyer and Leard), or where local custom and ideology upholding the rules is widely supported, as in Buen Hombre.

15. where state bureaucracies are not well funded or well developed and a community-based system will be more effective (Stoffle et al.; Dyer and Leard);

16. where nontraditional users are not dominated by elites or nonlocal interests that do not have to suffer the consequences of their actions directly (Ostrom; Jentoft and Mikalsen);

17. where interception can be controlled, or where locals are not overly affected by intercepting behavior of outside users (Felt; Jentoft and Mikalsen);

18. where mutual monitoring is possible and not too costly (Felt, Ostrom);

19. where government is used as an appeal mechanism to work out controversy, apply standards, or insure that basic conservation and equity standards are not violated (Stoffle et al.);

20. where there is a balance between co-management at the national level and at the natural resource community level.

Conclusion

Folk management and co-management fisheries regimes that have arisen around the world show promise for addressing many of the issues of sustainability and equity that plague fisheries management today. However, the development of such regimes is by no means automatic or simple, nor is their survival assured. This volume provides examples of impediments to the development of folk management, as well as instances of success. Drawing upon the rich case history material here and the larger multidisciplinary literature, this chapter proposed twenty working hypotheses about conditions under which folk management can either arise or persist in the late twentieth century.

Policy Lessons and Implications

Practical applications of folk knowledge for area-based planning and management face challenges to combining and integrating such knowledge into management by state scientists. These are as follows:

1. how to know and record traditions (where practices are unconscious, translate into conscious ones);

2. how to synthesize traditions/practices into a usable framework;

3. how to find appropriate unit(s) of analysis;

4. how to use folk knowledge to build an analysis of an ecosystem through integrating such knowledge into an areawide collection of data;

5. how to find ways to create annual monitoring and analysis that is systematic, valid, and reliable and to integrate questions raised by observations of traditional users;

6. how to find a way to communicate the foregoing effectively to biologists;

7. how to find a way to communicate biologists' concerns and questions to local observers and holders of traditional knowledge;

8. how to turn traditional knowledge into a teaching tool to train local managers who will work with state biologists;

9. how to find the right forum for the comfort level of local knowledge holders to communicate effectively with scientists or youth (reports indicate that elders become bored in scientific caucuses, that elders are uncomfortable in classroom settings, that the most comfortable teaching environment is outdoors, and that the most effective teaching method is teaching by doing, with observation by those being taught);

10. if it is a multi-stakeholder process, how to use planners skilled in interest-based mediation to educate communities about resource needs and to help the community arrive at a consensus.

References

Berkes, Fikret, ed.
  1989 Common Property Resources: Ecology and Community-Based Sustainable Development. New York: Columbia University Press.

Berkes, Fikret, R. Preston, and P. George
  1991 The Evolution in Theory and Practice of the Joint Administration of Living Resources. Alternatives 18(2):12–18.

Blomquist, William
  1992 Dividing the Waters: Governing Groundwater in Southern California. San Francisco: ICS Press.

Feit, Harvey
    1988   Self-Management and State-Management: Forms of Knowing and Man-
            aging Northern Wildlife. Pp. 72–91 in M. Freeman and L. Carbyn, eds.,
            Traditional Knowledge and Renewable Resource Management in North-
            ern Regions. Edmonton: University of Alberta Press.

Freeman, Milton M.R.
    1989   The Alaska Eskimo Whaling Commission: Successful Co-Management
            Under Extreme Conditions. Pp. 154–169 in E. Pinkerton, ed., Co-Opera-
            tive Management of Local Fisheries: New Directions for Improving
            Management and Community Development. Vancouver: University of
            British Columbia Press.

Lowi, Theodore
    1969   The End of Liberalism. New York: Norton.

Martin, Kent
    1979   Play by the Rules or Don't Play at All: Space Division and Resource
            Allocation in a Rural Newfoundland Fishing Community. Pp. 277–298
            in Raoul Andersen, ed., North Atlantic Maritime Cultures: Anthropologi-
            cal Essays on Changing Adaptations. World Anthropology Series. The
            Hague, Netherlands: Mouton.

McCay, Bonnie J.
    1989   Co-Management of a Clam Revitalization Project: The New Jersey
            "Spawner Sanctuary" Program. Pp. 103–124 in E. Pinkerton, ed., Co-Op-
            erative Management of Local Fisheries: New Directions for Improving
            Management and Community Development. Vancouver: University of
            British Columbia Press.

McCay, Bonnie J., and James M. Acheson, eds.
    1987   The Question of the Commons: The Culture and Ecology of Communal
            Resources. Tucson: University of Arizona Press.

McFarland, Andrew
    1987   Interest Groups and Theories of Power in America. British Journal of
            Political Science 17:129–147.

Ostrom, Elinor
    1990   Governing the Commons: The Evolution of Institutions for Collective
            Action. Cambridge, England: Cambridge University Press.

Pinkerton, Evelyn
    1988   Co-Operative Management of Local Fisheries: A Route to Development.
            Pp. 257–271 in John Bennett and John Bowen, eds., Production and
            Autonomy: Anthropological Studies and Critiques of Development.
            Lanham, MD: University Press of America.
    1989   Attaining Better Fisheries Management Through Co-Management: Pros-
            pects, Problems, and Propositions. Pp. 3–36 in E. Pinkerton, ed., Co-Op-
            erative Management of Local Fisheries: New Directions for Improving
            Management and Community Development. Vancouver: University of
            British Columbia Press.

1992 Translating Legal Rights Into Management Practice: Overcoming Barriers to the Exercise of Co-Management. Human Organization 51(4):330–341.

Taylor, Lawrence
1987 "The River Would Run Red With Blood": Community and Common Property in an Irish Fishing Settlement. Pp. 290–310 in Bonnie J. McCay and James M. Acheson, eds., The Question of the Commons: The Culture and Ecology of Communal Resources. Tucson: University of Arizona Press.

Index

adaptation: and continuity of local knowledge, 174–77

advisory committees: co-management and fishing industry, 102

agriculture: fishing and seasonality of occupational activities in Dominican Republic, 125; pesticide and water quality issues, 46; and small-scale fishing in fjords of Norway, 290

Alabama: folk management in oyster fishery, 74–76, 77–78, 80–81

Alaska: *Exxon Valdez* oil spill and fishery co-management, 207–29

altruism: folk management and formal regulation in Maine lobster fishery, 245

Alyeska Pipeline Service Company, 219, 230n

animism: fish sacralization, 139

anthropology: incorporation of studies of fishing communities into fisheries management, 55–56, 91. *See also* maritime anthropology

Apalachicola Bay: oyster fishery in Florida, 71–73

apprenticeship: fishers and experiential learning in Dominican Republic, 118; local knowledge and training of fishers, 165

aquaculture: oyster mariculture in Louisiana, 66–67; rancho system at Lake Chapala, Mexico, 30–32; research projects on cod sea ranching in fjords of Norway, 299, 311

Asian Americans: oyster industry in Gulf states, 68–69, 75

Atlantic Salmon Advisory Board (ASAB - Canada), 259, 275

authority: accommodations of co-management to local/state government, 324; fishers and enforcement of regulations in Dominican Republic, 135; recognition of resource users' rights to organize at Lake Chapala, Mexico, 18–19, 28–34, 35–37

bait fishing: and depletion of stocks in Fiji, 185

biologists: assessment of stocks in Canadian salmon fishery, 271–74, 281, 283n; management of oyster fishery on Texas Gulf coast, 92–109; role in fisheries management, 92–93. *See also* biology

biology: cyclic anticipatory utilization model, 59–61; role of information on in regulation of Norwegian fjord fisheries, 306; stability and success of oyster fisheries in Gulf states, 82. *See also* biologists; ecology; ecosystems; environment

bioremediation: of oil spills, 219

boundary definition: folk practices in Maine lobster fishery, 239; open and closed natural resource communities, 61–62; study of resource management at Lake Chapala, Mexico, 18–19, 21–28, 33, 34–37

Brazil: knowledge transmission among fishers, 165

California: groundwater users' associations as interest groups, 330; institutional failure in fisheries management, 149

Canada: folk management as model for formal regulation in Newfoundland lobster fishery, 237–47; social construction of indigenous knowledge among Atlantic salmon fishers, 252–83. *See also* Native Americans; Pacific Northwest

Caribbean: folk management and conservation ethics among fishers in Dominican Republic, 115–36; local knowledge of fish behavior, 177–78

China: fish sacralization, 145–48

closure: in community-based traditional marine resource-management systems, 187; management of oyster fish-